FREE Test Taking Tips DVD Offer

To help us better serve you, we have developed a Test Taking Tips DVD that we would like to give you for FREE. **This DVD covers world-class test taking tips that you can use to be even more successful when you are taking your test.**

All that we ask is that you email us your feedback about your study guide. Please let us know what you thought about it – whether that is good, bad or indifferent.

To get your **FREE Test Taking Tips DVD**, email freedvd@studyguideteam.com with "FREE DVD" in the subject line and the following information in the body of the email:

   a. The title of your study guide.

   b. Your product rating on a scale of 1-5, with 5 being the highest rating.

   c. Your feedback about the study guide. What did you think of it?

   d. Your full name and shipping address to send your free DVD.

If you have any questions or concerns, please don't hesitate to contact us at freedvd@studyguideteam.com.

Thanks again!

ACT Prep Questions Book 2021-2022

3 ACT Practice Tests with Detailed Answer Explanations [4th Edition]

Joshua Rueda

Written and edited by TPB Publishing.

TPB Publishing is not associated with or endorsed by any official testing organization. TPB Publishing is a publisher of unofficial educational products. All test and organization names are trademarks of their respective owners. Content in this book is included for utilitarian purposes only and does not constitute an endorsement by TPB Publishing of any particular point of view.

Interested in buying more than 10 copies of our product? Contact us about bulk discounts:
bulkorders@studyguideteam.com

ISBN 13: 9781637752487
ISBN 10: 1637752482

Table of Contents

Quick Overview

As you draw closer to taking your exam, effective preparation becomes more and more important. Thankfully, you have this study guide to help you get ready. Use this guide to help keep your studying on track and refer to it often.

This study guide contains several key sections that will help you be successful on your exam. The guide contains tips for what you should do the night before and the day of the test. Also included are test-taking tips. Knowing the right information is not always enough. Many well-prepared test takers struggle with exams. These tips will help equip you to accurately read, assess, and answer test questions.

A large part of the guide is devoted to showing you what content to expect on the exam and to helping you better understand that content. In this guide are practice test questions so that you can see how well you have grasped the content. Then, answer explanations are provided so that you can understand why you missed certain questions.

Don't try to cram the night before you take your exam. This is not a wise strategy for a few reasons. First, your retention of the information will be low. Your time would be better used by reviewing information you already know rather than trying to learn a lot of new information. Second, you will likely become stressed as you try to gain a large amount of knowledge in a short amount of time. Third, you will be depriving yourself of sleep. So be sure to go to bed at a reasonable time the night before. Being well-rested helps you focus and remain calm.

Be sure to eat a substantial breakfast the morning of the exam. If you are taking the exam in the afternoon, be sure to have a good lunch as well. Being hungry is distracting and can make it difficult to focus. You have hopefully spent lots of time preparing for the exam. Don't let an empty stomach get in the way of success!

When travelling to the testing center, leave earlier than needed. That way, you have a buffer in case you experience any delays. This will help you remain calm and will keep you from missing your appointment time at the testing center.

Be sure to pace yourself during the exam. Don't try to rush through the exam. There is no need to risk performing poorly on the exam just so you can leave the testing center early. Allow yourself to use all of the allotted time if needed.

Remain positive while taking the exam even if you feel like you are performing poorly. Thinking about the content you should have mastered will not help you perform better on the exam.

Once the exam is complete, take some time to relax. Even if you feel that you need to take the exam again, you will be well served by some down time before you begin studying again. It's often easier to convince yourself to study if you know that it will come with a reward!

Test-Taking Strategies

1. Predicting the Answer

When you feel confident in your preparation for a multiple-choice test, try predicting the answer before reading the answer choices. This is especially useful on questions that test objective factual knowledge. By predicting the answer before reading the available choices, you eliminate the possibility that you will be distracted or led astray by an incorrect answer choice. You will feel more confident in your selection if you read the question, predict the answer, and then find your prediction among the answer choices. After using this strategy, be sure to still read all of the answer choices carefully and completely. If you feel unprepared, you should not attempt to predict the answers. This would be a waste of time and an opportunity for your mind to wander in the wrong direction.

2. Reading the Whole Question

Too often, test takers scan a multiple-choice question, recognize a few familiar words, and immediately jump to the answer choices. Test authors are aware of this common impatience, and they will sometimes prey upon it. For instance, a test author might subtly turn the question into a negative, or he or she might redirect the focus of the question right at the end. The only way to avoid falling into these traps is to read the entirety of the question carefully before reading the answer choices.

3. Looking for Wrong Answers

Long and complicated multiple-choice questions can be intimidating. One way to simplify a difficult multiple-choice question is to eliminate all of the answer choices that are clearly wrong. In most sets of answers, there will be at least one selection that can be dismissed right away. If the test is administered on paper, the test taker could draw a line through it to indicate that it may be ignored; otherwise, the test taker will have to perform this operation mentally or on scratch paper. In either case, once the obviously incorrect answers have been eliminated, the remaining choices may be considered. Sometimes identifying the clearly wrong answers will give the test taker some information about the correct answer. For instance, if one of the remaining answer choices is a direct opposite of one of the eliminated answer choices, it may well be the correct answer. The opposite of obviously wrong is obviously right! Of course, this is not always the case. Some answers are obviously incorrect simply because they are irrelevant to the question being asked. Still, identifying and eliminating some incorrect answer choices is a good way to simplify a multiple-choice question.

4. Don't Overanalyze

Anxious test takers often overanalyze questions. When you are nervous, your brain will often run wild, causing you to make associations and discover clues that don't actually exist. If you feel that this may be a problem for you, do whatever you can to slow down during the test. Try taking a deep breath or counting to ten. As you read and consider the question, restrict yourself to the particular words used by the author. Avoid thought tangents about what the author *really* meant, or what he or she was *trying* to say. The only things that matter on a multiple-choice test are the words that are actually in the question. You must avoid reading too much into a multiple-choice question, or supposing that the writer meant something other than what he or she wrote.

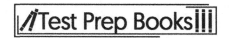

5. No Need for Panic

It is wise to learn as many strategies as possible before taking a multiple-choice test, but it is likely that you will come across a few questions for which you simply don't know the answer. In this situation, avoid panicking. Because most multiple-choice tests include dozens of questions, the relative value of a single wrong answer is small. As much as possible, you should compartmentalize each question on a multiple-choice test. In other words, you should not allow your feelings about one question to affect your success on the others. When you find a question that you either don't understand or don't know how to answer, just take a deep breath and do your best. Read the entire question slowly and carefully. Try rephrasing the question a couple of different ways. Then, read all of the answer choices carefully. After eliminating obviously wrong answers, make a selection and move on to the next question.

6. Confusing Answer Choices

When working on a difficult multiple-choice question, there may be a tendency to focus on the answer choices that are the easiest to understand. Many people, whether consciously or not, gravitate to the answer choices that require the least concentration, knowledge, and memory. This is a mistake. When you come across an answer choice that is confusing, you should give it extra attention. A question might be confusing because you do not know the subject matter to which it refers. If this is the case, don't eliminate the answer before you have affirmatively settled on another. When you come across an answer choice of this type, set it aside as you look at the remaining choices. If you can confidently assert that one of the other choices is correct, you can leave the confusing answer aside. Otherwise, you will need to take a moment to try to better understand the confusing answer choice. Rephrasing is one way to tease out the sense of a confusing answer choice.

7. Your First Instinct

Many people struggle with multiple-choice tests because they overthink the questions. If you have studied sufficiently for the test, you should be prepared to trust your first instinct once you have carefully and completely read the question and all of the answer choices. There is a great deal of research suggesting that the mind can come to the correct conclusion very quickly once it has obtained all of the relevant information. At times, it may seem to you as if your intuition is working faster even than your reasoning mind. This may in fact be true. The knowledge you obtain while studying may be retrieved from your subconscious before you have a chance to work out the associations that support it. Verify your instinct by working out the reasons that it should be trusted.

8. Key Words

Many test takers struggle with multiple-choice questions because they have poor reading comprehension skills. Quickly reading and understanding a multiple-choice question requires a mixture of skill and experience. To help with this, try jotting down a few key words and phrases on a piece of scrap paper. Doing this concentrates the process of reading and forces the mind to weigh the relative importance of the question's parts. In selecting words and phrases to write down, the test taker thinks about the question more deeply and carefully. This is especially true for multiple-choice questions that are preceded by a long prompt.

9. Subtle Negatives

One of the oldest tricks in the multiple-choice test writer's book is to subtly reverse the meaning of a question with a word like *not* or *except*. If you are not paying attention to each word in the question, you can easily be led astray by this trick. For instance, a common question format is, "Which of the following is...?" Obviously, if the question instead is, "Which of the following is not...?," then the answer will be quite different. Even worse, the test makers are aware of the potential for this mistake and will include one answer choice that would be correct if the question were not negated or reversed. A test taker who misses the reversal will find what he or she believes to be a correct answer and will be so confident that he or she will fail to reread the question and discover the original error. The only way to avoid this is to practice a wide variety of multiple-choice questions and to pay close attention to each and every word.

10. Reading Every Answer Choice

It may seem obvious, but you should always read every one of the answer choices! Too many test takers fall into the habit of scanning the question and assuming that they understand the question because they recognize a few key words. From there, they pick the first answer choice that answers the question they believe they have read. Test takers who read all of the answer choices might discover that one of the latter answer choices is actually *more* correct. Moreover, reading all of the answer choices can remind you of facts related to the question that can help you arrive at the correct answer. Sometimes, a misstatement or incorrect detail in one of the latter answer choices will trigger your memory of the subject and will enable you to find the right answer. Failing to read all of the answer choices is like not reading all of the items on a restaurant menu: you might miss out on the perfect choice.

11. Spot the Hedges

One of the keys to success on multiple-choice tests is paying close attention to every word. This is never truer than with words like almost, most, some, and sometimes. These words are called "hedges" because they indicate that a statement is not totally true or not true in every place and time. An absolute statement will contain no hedges, but in many subjects, the answers are not always straightforward or absolute. There are always exceptions to the rules in these subjects. For this reason, you should favor those multiple-choice questions that contain hedging language. The presence of qualifying words indicates that the author is taking special care with his or her words, which is certainly important when composing the right answer. After all, there are many ways to be wrong, but there is only one way to be right! For this reason, it is wise to avoid answers that are absolute when taking a multiple-choice test. An absolute answer is one that says things are either all one way or all another. They often include words like *every*, *always*, *best*, and *never*. If you are taking a multiple-choice test in a subject that doesn't lend itself to absolute answers, be on your guard if you see any of these words.

12. Long Answers

In many subject areas, the answers are not simple. As already mentioned, the right answer often requires hedges. Another common feature of the answers to a complex or subjective question are qualifying clauses, which are groups of words that subtly modify the meaning of the sentence. If the question or answer choice describes a rule to which there are exceptions or the subject matter is complicated, ambiguous, or confusing, the correct answer will require many words in order to be expressed clearly and accurately. In essence, you should not be deterred by answer choices that seem excessively long. Oftentimes, the author of the text will not be able to write the correct answer without

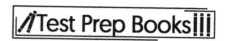

offering some qualifications and modifications. Your job is to read the answer choices thoroughly and completely and to select the one that most accurately and precisely answers the question.

13. Restating to Understand

Sometimes, a question on a multiple-choice test is difficult not because of what it asks but because of how it is written. If this is the case, restate the question or answer choice in different words. This process serves a couple of important purposes. First, it forces you to concentrate on the core of the question. In order to rephrase the question accurately, you have to understand it well. Rephrasing the question will concentrate your mind on the key words and ideas. Second, it will present the information to your mind in a fresh way. This process may trigger your memory and render some useful scrap of information picked up while studying.

14. True Statements

Sometimes an answer choice will be true in itself, but it does not answer the question. This is one of the main reasons why it is essential to read the question carefully and completely before proceeding to the answer choices. Too often, test takers skip ahead to the answer choices and look for true statements. Having found one of these, they are content to select it without reference to the question above. Obviously, this provides an easy way for test makers to play tricks. The savvy test taker will always read the entire question before turning to the answer choices. Then, having settled on a correct answer choice, he or she will refer to the original question and ensure that the selected answer is relevant. The mistake of choosing a correct-but-irrelevant answer choice is especially common on questions related to specific pieces of objective knowledge. A prepared test taker will have a wealth of factual knowledge at his or her disposal, and should not be careless in its application.

15. No Patterns

One of the more dangerous ideas that circulates about multiple-choice tests is that the correct answers tend to fall into patterns. These erroneous ideas range from a belief that B and C are the most common right answers, to the idea that an unprepared test-taker should answer "A-B-A-C-A-D-A-B-A." It cannot be emphasized enough that pattern-seeking of this type is exactly the WRONG way to approach a multiple-choice test. To begin with, it is highly unlikely that the test maker will plot the correct answers according to some predetermined pattern. The questions are scrambled and delivered in a random order. Furthermore, even if the test maker was following a pattern in the assignation of correct answers, there is no reason why the test taker would know which pattern he or she was using. Any attempt to discern a pattern in the answer choices is a waste of time and a distraction from the real work of taking the test. A test taker would be much better served by extra preparation before the test than by reliance on a pattern in the answers.

FREE DVD OFFER

Don't forget that doing well on your exam includes both understanding the test content and understanding how to use what you know to do well on the test. We offer a completely FREE Test Taking Tips DVD that covers world class test taking tips that you can use to be even more successful when you are taking your test.

All that we ask is that you email us your feedback about your study guide. To get your **FREE Test Taking Tips DVD**, email freedvd@studyguideteam.com with "FREE DVD" in the subject line and the following information in the body of the email:

- The title of your study guide.
- Your product rating on a scale of 1-5, with 5 being the highest rating.
- Your feedback about the study guide. What did you think of it?
- Your full name and shipping address to send your free DVD.

Introduction to the ACT

Function of the Test

The ACT is one of two national standardized college entrance examinations (the SAT being the other). Most prospective college students take the ACT or the SAT, and it is increasingly common for students to take both. All four-year colleges and universities in the United States accept the ACT for admissions purposes, and some require it. Some of those schools also use ACT subject scores for placement purposes. Sixteen states also require all high school juniors to take the ACT as part of the states' school evaluation efforts.

The vast majority of people taking the ACT are high school juniors and seniors who intend to apply to college. Traditionally, the SAT was more commonly taken than the ACT, particularly among students on the East and West coasts. However, the popularity of the ACT has grown dramatically in recent years and is now commonly taken by students in all fifty states. In fact, starting in 2013, more test takers took the ACT than the SAT. In 2015, 1.92 million students took the ACT. About 28 percent of 2015 high school graduates taking the ACT met the test's college-readiness benchmarks in all four subjects, while 31 percent met none of the benchmarks.

Test Administration

The ACT is offered on six dates throughout the year in the U.S. and Canada, and on five of those same dates in other countries. The registration fee includes score reports for four colleges, with additional reports available for purchase. There is a separate registration fee for the optional writing section.

On test dates, the ACT is administered at test centers throughout the world. The test centers are usually high schools or colleges, with several locations usually available in significant population centers.

Test takers can retake the ACT as frequently as the test is offered, up to a maximum of twelve times; although, individual colleges may have limits on how many retakes they will consider. Starting September 2020, test takers can opt to take individual subsections during a retake attempt rather than needing to retake the entire exam.

The ACT does provide reasonable accommodations to test takers with professionally-documented disabilities.

Test Format

The ACT consists of 215 multiple-choice questions in four subject areas (English, mathematics, reading, and science) and takes about three hours and thirty minutes to complete. It also has an optional writing test, which takes an additional forty minutes.

The English section is 45 minutes long and contains 75 questions on usage, language mechanics, and rhetorical skills. The Mathematics section is 60 minutes long and contains 60 questions on algebra, geometry, and elementary trigonometry. Calculators that meet the ACT's calculator policy are permitted on the Mathematics section. The Reading section is 35 minutes long and contains four written passages with ten questions per passage. The Science section is 35 minutes long and contains 40 questions.

The Writing section is forty minutes long and is always given at the end so that test takers not wishing to take it may leave after completing the other four sections. This section consists of one essay in which students must analyze three different perspectives on a broad social issue. Although the Writing section is optional, some colleges do require it.

Section	Length	Questions
English	45 minutes	75
Mathematics	60 minutes	60
Reading	35 minutes	40
Science	35 minutes	40
Writing (optional)	40 minutes	1 essay

Scoring

Test takers receive a score between 1 and 36 for each of the four subject areas. Those scores are averaged together to give a Composite Score, which is the primary score reported as an "ACT score." The most prestigious schools typically admit students with Composite ACT Scores in the low 30's. Other selective schools typically admit students with scores in the high 20's. Traditional colleges more likely admit students with scores in the low 20's, while community colleges and other more open schools typically accept students with scores in the high teens. In 2015, the average composite score among all test takers (including those not applying to college) was 21.

Before September 2020, any test takers who opted to retake the ACT had to retake the entire exam; moreover, the composite score was calculated from the scores achieved over a single testing session. As of September 2020, test takers can select individual sections to retake. Additionally, a "superscore" will be calculated, and will reflect the test taker's highest scores on each of the subsections from any of their attempts at the test. For example, if the test taker took the test twice and earned a higher score on Math and English during the first administration, and a higher score on Reading, Science, and Writing the second time, the "superscore" would be the sum of these top performances.

ACT Practice Test #1

English

Questions 1–9 are based on the following passage:

While all dogs (1) <u>descend through gray wolves</u>, it's easy to notice that dog breeds come in a variety of shapes and sizes. With such a (2) <u>drastic range of traits, appearances and body types</u> dogs are one of the most variable and adaptable species on the planet. (3) <u>But why so many differences.</u> The answer is that humans have actually played a major role in altering the biology of dogs. (4) <u>This was done through a process called selective breeding.</u>

(5) <u>Selective breeding which is also called artificial selection is the processes</u> in which animals with desired traits are bred in order to produce offspring that share the same traits. In natural selection, (6) <u>animals must adapt to their environments increase their chance of survival</u>. Over time, certain traits develop in animals that enable them to thrive in these environments. Those animals with more of these traits, or better versions of these traits, gain an (7) <u>advantage over others of their species.</u> Therefore, the animal's chances to mate are increased and these useful (8) <u>genes are passed into their offspring.</u> With dog breeding, humans select traits that are desired and encourage more of these desired traits in other dogs by breeding dogs that already have them.

The reason for different breeds of dogs is that there were specific needs that humans wanted to fill with their animals. For example, scent hounds are known for their extraordinary ability to track game through scent. These breeds are also known for their endurance in seeking deer and other prey. Therefore, early hunters took dogs that displayed these abilities and bred them to encourage these traits. Later, these generations took on characteristics that aided these desired traits. (9) <u>For example, Bloodhounds</u> have broad snouts and droopy ears that fall to the ground when they smell. These physical qualities not only define the look of the Bloodhound, but also contribute to their amazing tracking ability. The broad snout is able to define and hold onto scents longer than many other breeds. The long, floppy ears serve to collect and hold the scents the earth holds so that the smells are clearer and able to be distinguished.

1. Which of the following would be the best choice for this sentence (reproduced below)?

   While all dogs (1) <u>descend through gray wolves</u>, it's easy to notice that dog breeds come in a variety of shapes and sizes.

   a. NO CHANGE
   b. descend by gray wolves
   c. descend from gray wolves
   d. descended through gray wolves

2. Which of the following would be the best choice for this sentence (reproduced below)?

With such a (2) <u>drastic range of traits, appearances and body types</u>, dogs are one of the most variable and adaptable species on the planet.

a. NO CHANGE
b. drastic range of traits, appearances, and body types,
c. drastic range of traits and appearances and body types,
d. drastic range of traits, appearances, as well as body types,

3. Which of the following would be the best choice for this sentence (reproduced below)?

(3) <u>But why so many differences.</u>

a. NO CHANGE
b. But are there so many differences?
c. But why so many differences are there.
d. But why so many differences?

4. Which of the following would be the best choice for this sentence (reproduced below)?

(4) <u>This was done through a process called selective breeding.</u>

a. NO CHANGE
b. This was done, through a process called selective breeding.
c. This was done, through a process, called selective breeding.
d. This was done through selective breeding, a process.

5. Which of the following would be the best choice for this sentence (reproduced below)?

(5) <u>Selective breeding which is also called artificial selection is the processes</u> in which animals with desired traits are bred in order to produce offspring that share the same traits.

a. NO CHANGE
b. Selective breeding, which is also called artificial selection is the processes
c. Selective breeding which is also called, artificial selection, is the processes
d. Selective breeding, which is also called artificial selection, is the processes

6. Which of the following would be the best choice for this sentence (reproduced below)?

In natural selection, (6) animals must adapt to their environments increase their chance of survival.

a. NO CHANGE
b. animals must adapt to their environments to increase their chance of survival.
c. animals must adapt to their environments, increase their chance of survival.
d. animals must adapt to their environments, increasing their chance of survival.

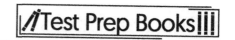

7. Which of the following would be the best choice for this sentence (reproduced below)?

Those animals with more of these traits, or better versions of these traits, gain an (7) advantage over others of their species.

    a. NO CHANGE
    b. advantage over others, of their species.
    c. advantages over others of their species.
    d. advantage over others.

8. Which of the following would be the best choice for this sentence (reproduced below)?

Therefore, the animal's chances to mate are increased and these useful (8) genes are passed into their offspring.

    a. NO CHANGE
    b. genes are passed onto their offspring.
    c. genes are passed on to their offspring.
    d. genes are passed within their offspring.

9. Which of the following would be the best choice for this sentence (reproduced below)?

(9) For example, Bloodhounds have broad snouts and droopy ears that fall to the ground when they smell.

    a. NO CHANGE
    b. For example, Bloodhounds,
    c. For example Bloodhounds
    d. For example, bloodhounds

Questions 10–18 are based on the following passage:

I'm not alone when I say that it's hard to pay attention sometimes. I can't count how many times I've sat in a classroom, lecture, speech, or workshop and (10) been bored to tears or rather sleep. (11) Usually I turn to doodling in order to keep awake. This never really helps; I'm not much of an artist. Therefore, after giving up on drawing a masterpiece, I would just concentrate on keeping my eyes open and trying to be attentive. This didn't always work because I wasn't engaged in what was going on.

(12) Sometimes in particularly dull seminars, I'd imagine comical things going on in the room or with the people trapped in the room with me. Why? (13) Because I wasn't invested in what was going on I wasn't motivated to listen. I'm not going to write about how I conquered the difficult task of actually paying attention in a difficult or unappealing class—it can be done, sure. I have sat through the very epitome of boredom (in my view at least) several times and come away learning something. (14) Everyone probably has had to at one time do this. What I want to talk about is that profound moment when curiosity is sparked (15) in another person drawing them to pay attention to what is before them and expand their knowledge.

What really makes people pay attention? (16) Easy it's interest. This doesn't necessarily mean (17) embellishing subject matter drawing people's attention. This won't always work. However,

11

an individual can present material in a way that is clear to understand and actually engages the audience. Asking questions to the audience or class will make them a part of the topic at hand. Discussions that make people think about the content and (18) <u>how it applies to there lives world and future is key.</u> If math is being discussed, an instructor can explain the purpose behind the equations or perhaps use real-world applications to show how relevant the topic is. When discussing history, a lecturer can prompt students to imagine themselves in the place of key figures and ask how they might respond. The bottom line is to explore the ideas rather than just lecture. Give people the chance to explore material from multiple angles, and they'll be hungry to keep paying attention for more information.

10. Which of the following would be the best choice for this sentence (reproduced below)?

I can't count how many times I've sat in a classroom, lecture, speech, or workshop and (10) <u>been bored to tears or rather sleep.</u>

a. NO CHANGE
b. been bored to, tears, or rather sleep.
c. been bored, to tears or rather sleep.
d. been bored to tears or, rather, sleep.

11. Which of the following would be the best choice for this sentence (reproduced below)?

(11) <u>Usually I turn to doodling in order to keep awake.</u>

a. NO CHANGE
b. Usually, I turn to doodling in order to keep awake.
c. Usually I turn to doodling, in order, to keep awake.
d. Usually I turned to doodling in order to keep awake.

12. Which of the following would be the best choice for this sentence (reproduced below)?

(12) <u>Sometimes in particularly dull seminars,</u> I'd imagine comical things going on in the room or with the people trapped in the room with me.

a. NO CHANGE
b. Sometimes, in particularly, dull seminars,
c. Sometimes in particularly dull seminars
d. Sometimes in particularly, dull seminars,

13. Which of the following would be the best choice for this sentence (reproduced below)?

(13) <u>Because I wasn't invested in what was going on I wasn't motivated to listen.</u>

a. NO CHANGE
b. Because I wasn't invested, in what was going on, I wasn't motivated to listen.
c. Because I wasn't invested in what was going on. I wasn't motivated to listen.
d. I wasn't motivated to listen because I wasn't invested in what was going on.

14. Which of the following would be the best choice for this sentence (reproduced below)?

(14) Everyone probably has had to at one time do this.

a. NO CHANGE
b. Everyone probably has had to, at one time. Do this.
c. Everyone's probably had to do this at some time.
d. At one time everyone probably has had to do this.

15. Which of the following would be the best choice for this sentence (reproduced below)?

What I want to talk about is that profound moment when curiosity is sparked (15) in another person drawing them to pay attention to what is before them and expand their knowledge.

a. NO CHANGE
b. in another person, drawing them to pay attention
c. in another person; drawing them to pay attention to what is before them.
d. in another person, drawing them to pay attention to what is before them.

16. Which of the following would be the best choice for this sentence (reproduced below)?

(16) Easy it's interest.

a. NO CHANGE
b. Easy it is interest.
c. Easy. It's interest.
d. Easy—it's interest.

17. Which of the following would be the best choice for this sentence (reproduced below)?

This doesn't necessarily mean (17) embellishing subject matter drawing people's attention.

a. NO CHANGE
b. embellishing subject matter which draws people's attention.
c. embellishing subject matter to draw people's attention.
d. embellishing subject matter for the purpose of drawing people's attention.

18. Which of the following would be the best choice for this sentence (reproduced below)?

Discussions that make people think about the content and (18) how it applies to there lives world and future is key.

a. NO CHANGE
b. how it applies to their lives, world, and future is key.
c. how it applied to there lives world and future is key.
d. how it applies to their lives, world and future is key.

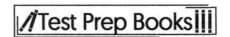

Questions 19–27 are based on the following passage:

Since the first discovery of dinosaur bones, (19) <u>scientists has made strides in technological development and methodologies used to investigate</u> these extinct animals. We know more about dinosaurs than ever before and are still learning fascinating new things about how they looked and lived. However, one has to ask, (20) <u>how if earlier perceptions of dinosaurs</u> continue to influence people's understanding of these creatures? Can these perceptions inhibit progress towards further understanding of dinosaurs?

(21) <u>The biggest problem with studying dinosaurs is simply that there are no living dinosaurs to observe.</u> All discoveries associated with these animals are based on physical remains. To gauge behavioral characteristics, scientists cross-examine these (22) <u>finds with living animals that seem similar in order to gain understanding.</u> While this method is effective, these are still deductions. Some ideas about dinosaurs can't be tested and confirmed simply because humans can't replicate a living dinosaur. For example, a Spinosaurus has a large sail, or a finlike structure that grows from its back. Paleontologists know this sail exists and have ideas for the function of (23) <u>the sail however they are uncertain of which idea is the true function.</u> Some scientists believe (24) <u>the sail serves to regulate the Spinosaurus' body temperature and yet others believe its used to attract mates.</u> Still, other scientists think the sail is used to intimidate other predatory dinosaurs for self-defense. These are all viable explanations, but they are also influenced by what scientists know about modern animals. (25) <u>Yet, it's quite possible</u> that the sail could hold a completely unique function.

While it's (26) <u>plausible, even likely that dinosaurs share many</u> traits with modern animals, there is the danger of overattributing these qualities to a unique, extinct species. For much of the early nineteenth century, when people first started studying dinosaur bones, the assumption was that they were simply giant lizards. (27) <u>For the longest time this image was the prevailing view on dinosaurs,</u> until evidence indicated that they were more likely warm blooded. Scientists have also discovered that many dinosaurs had feathers and actually share many traits with modern birds.

19. Which of the following would be the best choice for this sentence (reproduced below)?

Since the first discovery of dinosaur bones, (19) <u>scientists has made strides in technological development and methodologies used to investigate</u> these extinct animals.

a. NO CHANGE
b. scientists has made strides in technological development, and methodologies, used to investigate
c. scientists have made strides in technological development and methodologies used to investigate
d. scientists, have made strides in technological development and methodologies used, to investigate

20. Which of the following would be the best choice for this sentence (reproduced below)?

However, one has to ask, (20) how if earlier perceptions of dinosaurs continue to influence people's understanding of these creatures?

a. NO CHANGE
b. how perceptions of dinosaurs
c. how, if, earlier perceptions of dinosaurs
d. whether earlier perceptions of dinosaurs

21. Which of the following would be the best choice for this sentence (reproduced below)?

(21) The biggest problem with studying dinosaurs is simply that there are no living dinosaurs to observe.

a. NO CHANGE
b. The biggest problem with studying dinosaurs is simple, that there are no living dinosaurs to observe.
c. The biggest problem with studying dinosaurs is simple. There are no living dinosaurs to observe.
d. The biggest problem with studying dinosaurs, is simply that there are no living dinosaurs to observe.

22. Which of the following would be the best choice for this sentence (reproduced below)?

To gauge behavioral characteristics, scientists cross-examine these (22) finds with living animals that seem similar in order to gain understanding.

a. NO CHANGE
b. finds with living animals to explore potential similarities.
c. finds with living animals to gain understanding of similarities.
d. finds with living animals that seem similar, in order, to gain understanding.

23. Which of the following would be the best choice for this sentence (reproduced below)?

Paleontologists know this sail exists and have ideas for the function of (23) the sail however they are uncertain of which idea is the true function.

a. NO CHANGE
b. the sail however, they are uncertain of which idea is the true function.
c. the sail however they are, uncertain, of which idea is the true function.
d. the sail; however, they are uncertain of which idea is the true function.

ACT Practice Test #1

24. Which of the following would be the best choice for this sentence (reproduced below)?

Some scientists believe (24) the sail serves to regulate the Spinosaurus' body temperature and yet others believe its used to attract mates.

a. NO CHANGE
b. the sail serves to regulate the Spinosaurus' body temperature, yet others believe it's used to attract mates.
c. the sail serves to regulate the Spinosaurus' body temperature and yet others believe it's used to attract mates.
d. the sail serves to regulate the Spinosaurus' body temperature however others believe it's used to attract mates.

25. Which of the following would be the best choice for this sentence (reproduced below)?

(25) Yet, it's quite possible that the sail could hold a completely unique function.

a. NO CHANGE
b. Yet, it's quite possible,
c. It's quite possible,
d. Its quite possible

26. Which of the following would be the best choice for this sentence (reproduced below)?

While it's (26) plausible, even likely that dinosaurs share many traits with modern animals, there is the danger of over attributing these qualities to a unique, extinct species.

a. NO CHANGE
b. plausible, even likely that, dinosaurs share many
c. plausible, even likely, that dinosaurs share many
d. plausible even likely that dinosaurs share many

27. Which of the following would be the best choice for this sentence (reproduced below)?

(27) For the longest time this image was the prevailing view on dinosaurs, until evidence indicated that they were more likely warm blooded.

a. NO CHANGE
b. For the longest time this was the prevailing view on dinosaurs
c. For the longest time, this image, was the prevailing view on dinosaurs
d. For the longest time this was the prevailing image of dinosaurs

Questions 28–36 are based on the following passage:

Everyone has heard the (28) idea of the end justifying the means; that would be Weston's philosophy. Weston is willing to cross any line, commit any act no matter how heinous, to achieve success in his goal. (29) Ransom is reviled by this fact, seeing total evil in Weston's plan. To do an evil act in order (30) to gain a result that's supposedly good would ultimately warp the final act. (31) This opposing viewpoints immediately distinguishes Ransom as the hero. In the conflict with Un-man, Ransom remains true to his moral principles, someone who refuses to be

16

compromised by power. Instead, Ransom makes it clear that by allowing such processes as murder and lying dictate how one attains a positive outcome, (32) <u>the righteous goal becomes corrupted.</u> The good end would not be truly good, but a twisted end that conceals corrupt deeds.

(33) <u>This idea of allowing necessary evils to happen, is very tempting, it is what Weston fell prey to.</u> (34) <u>The temptation of the evil spirit Un-man ultimately takes over Weston and he is possessed.</u> However, Ransom does not give into temptation. He remains faithful to the truth of what is right and incorrect. This leads him to directly face Un-man for the fate of Perelandra and its inhabitants.

Just as Weston was corrupted by the Un-man, (35) <u>Un-man after this seeks to tempt the Queen of Perelandra</u> to darkness. Ransom must literally (36) <u>show her the right path, to accomplish this, he does this based on the same principle as the "means to an end" argument</u>—that good follows good, and evil follows evil. Later in the plot, Weston/Un-man seeks to use deceptive reasoning to turn the queen to sin, pushing the queen to essentially ignore Melildil's rule to satisfy her own curiosity. In this sense, Un-man takes on the role of a false prophet, a tempter. Ransom must shed light on the truth, but this is difficult; his adversary is very clever and uses brilliant language. Ransom's lack of refinement heightens the weight of Un-man's corrupted logic, and so the Queen herself is intrigued by his logic.

Based on an excerpt from Perelandra *by C.S. Lewis*

28. Which of the following would be the best choice for this sentence (reproduced below)?

Everyone has heard the (28) <u>idea of the end justifying the means; that would be Weston's philosophy.</u>

a. NO CHANGE
b. idea of the end justifying the means; this is Weston's philosophy.
c. idea of the end justifying the means, this is the philosophy of Weston
d. idea of the end justifying the means. That would be Weston's philosophy.

29. Which of the following would be the best choice for this sentence (reproduced below)?

(29) <u>Ransom is reviled by this fact, seeing total evil in Weston's plan.</u>

a. NO CHANGE
b. Ransom is reviled by this fact; seeing total evil in Weston's plan.
c. Ransom, is reviled by this fact, seeing total evil in Weston's plan.
d. Ransom reviled by this, sees total evil in Weston's plan.

30. Which of the following would be the best choice for this sentence (reproduced below)?

To do an evil act in order (30) <u>to gain a result that's supposedly good would ultimately warp the final</u> <u>act.</u>

a. NO CHANGE
b. for an outcome that's for a greater good would ultimately warp the final act.
c. to gain a final act would warp its goodness.
d. to achieve a positive outcome would ultimately warp the goodness of the final act.

31. Which of the following would be the best choice for this sentence (reproduced below)?

(31) <u>This opposing viewpoints immediately distinguishes Ransom as the hero.</u>

a. NO CHANGE
b. This opposing viewpoints immediately distinguishes Ransom, as the hero.
c. This opposing viewpoint immediately distinguishes Ransom as the hero.
d. Those opposing viewpoints immediately distinguishes Ransom as the hero.

32. Which of the following would be the best choice for this sentence (reproduced below)?

Instead, Ransom makes it clear that by allowing such processes as murder and lying dictate how one attains a positive outcome, (32) <u>the righteous goal becomes corrupted.</u>

a. NO CHANGE
b. the goal becomes corrupted and no longer righteous.
c. the righteous goal becomes, corrupted.
d. the goal becomes corrupted, when once it was righteous.

33. Which of the following would be the best choice for this sentence (reproduced below)?

(33) <u>This idea of allowing necessary evils to happen, is very tempting, it is what Weston fell prey to.</u>

a. NO CHANGE
b. This idea of allowing necessary evils to happen, is very tempting. This is what Weston fell prey to.
c. This idea, allowing necessary evils to happen, is very tempting, it is what Weston fell prey to.
d. This tempting idea of allowing necessary evils to happen is what Weston fell prey to.

34. Which of the following would be the best choice for this sentence (reproduced below)?

(34) <u>The temptation of the evil spirit Un-man ultimately takes over Weston and he is possessed.</u>

a. NO CHANGE
b. The temptation of the evil spirit Un-man ultimately takes over and possesses Weston.
c. Weston is possessed as a result of the temptation of the evil spirit Un-man ultimately, who takes over.
d. The temptation of the evil spirit Un-man takes over Weston and he is possessed ultimately.

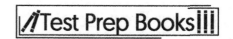

35. Which of the following would be the best choice for this sentence (reproduced below)?

Just as Weston was corrupted by the Un-man, (35) <u>Un-man after this seeks to tempt the Queen of Perelandra</u> to darkness.

a. NO CHANGE
b. Un-man, after this, would tempt the Queen of Perelandra
c. Un-man, after this, seeks to tempt the Queen of Perelandra
d. Un-man then seeks to tempt the Queen of Perelandra

36. Which of the following would be the best choice for this sentence (reproduced below)?

Ransom must literally (36) <u>show her the right path, to accomplish this, he does this based on the same principle as the "means to an end" argument</u>—that good follows good, and evil follows evil.

a. NO CHANGE
b. show her the right path. To accomplish this, he uses the same principle as the "means to an end" argument
c. show her the right path; to accomplish this he uses the same principle as the "means to an end" argument
d. show her the right path, to accomplish this, the same principle as the "means to an end" argument is applied

Questions 37–45 are based on the following passage:

(37) <u>What's clear about the news is today is that the broader the media</u> the more ways there are to tell a story. Even if different news groups cover the same story, individual newsrooms can interpret or depict the story differently than other counterparts. Stories can also change depending on the type of (38) <u>media in question incorporating different styles and unique</u> ways to approach the news. (39) <u>It is because of these respective media types that ethical and news-related subject matter can sometimes seem different or altered.</u> But how does this affect the narrative of the new story?

I began by investigating a written newspaper article from the Baltimore Sun. Instantly striking are the bolded headlines. (40) <u>These are clearly meant for direct the viewer</u> to the most exciting and important stories the paper has to offer. What was particularly noteworthy about this edition was that the first page dealt with two major ethical issues. (41) <u>On a national level there was a story</u> on the evolving Petraeus scandal involving his supposed affair. The other article was focused locally in Baltimore, a piece questioning the city's Ethics Board and their current director. Just as a television newscaster communicates the story through camera and dialogue, the printed article applies intentional and targeted written narrative style. More so than any of the mediums, a news article seems to be focused specifically on a given story without need to jump to another. Finer details are usually expanded on (42) <u>in written articles, usually people who</u> read newspapers or go online for web articles want more than a quick blurb. The diction of the story is also more precise and can be either straightforward or suggestive (43) <u>depending in earnest on the goal of the writer.</u> However, there's still plenty of room for opinions to be inserted into the text.

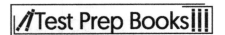

Usually, all news (44) <u>outlets have some sort of bias, it's just a question of how much</u> bias clouds the reporting. As long as this bias doesn't withhold information from the reader, it can be considered credible. (45) <u>However an over use of bias,</u> opinion, and suggestive language can rob readers of the chance to interpret the news events for themselves.

37. Which of the following would be the best choice for this sentence (reproduced below)?

(37) <u>What's clear about the news today is that the broader the media</u> the more ways there are to tell a story.

a. NO CHANGE
b. What's clear, about the news today, is that the broader the media
c. What's clear about today's news is that the broader the media
d. The news today is broader than earlier media

38. Which of the following would be the best choice for this sentence (reproduced below)?

Stories can also change depending on the type of (38) <u>media in question incorporating different styles and unique</u> ways to approach the news.

a. NO CHANGE
b. media in question; each incorporates unique styles and unique
c. media in question. To incorporate different styles and unique
d. media in question, incorporating different styles and unique

39. Which of the following would be the best choice for this sentence (reproduced below)?

(39) <u>It is because of these respective media types that ethical and news-related subject matter can sometimes seem different or altered.</u>

a. NO CHANGE
b. It is because of these respective media types, that ethical and news-related subject matter, can sometimes seem different or altered.
c. It is because of these respective media types, that ethical and news-related subject matter can sometimes seem different or altered.
d. It is because of these respective media types that ethical and news-related subject matter can sometimes seem different. Or altered.

40. Which of the following would be the best choice for this sentence (reproduced below)?

(40) <u>These are clearly meant for direct the viewer</u> to the most exciting and important stories the paper has to offer.

a. NO CHANGE
b. These are clearly meant for the purpose of giving direction to the viewer
c. These are clearly meant to direct the viewer
d. These are clearly meant for the viewer to be directed

41. Which of the following would be the best choice for this sentence (reproduced below)?

(41) On a national level there was a story on the evolving Petraeus scandal involving his supposed affair.

a. NO CHANGE
b. On a national level a story was there
c. On a national level; there was a story
d. On a national level, there was a story

42. Which of the following would be the best choice for this sentence (reproduced below)?

Finer details are usually expanded on (42) in written articles, usually people who read newspapers or go online for web articles want more than a quick blurb.

a. NO CHANGE
b. in written articles. People who usually
c. in written articles, usually, people who
d. in written articles usually people who

43. Which of the following would be the best choice for this sentence (reproduced below)?

The diction of the story is also more precise and can be either straightforward or suggestive (43) depending in earnest on the goal of the writer.

a. NO CHANGE
b. depending; in earnest on the goal of the writer.
c. depending, in earnest, on the goal of the writer.
d. the goal of the writer, in earnest, depends on the goal of the writer.

44. Which of the following would be the best choice for this sentence (reproduced below)?

Usually, all news (44) outlets have some sort of bias, it's just a question of how much bias clouds the reporting.

a. NO CHANGE
b. outlets have some sort of bias. Just a question of how much
c. outlets have some sort of bias it can just be a question of how much
d. outlets have some sort of bias, its just a question of how much

45. Which of the following would be the best choice for this sentence (reproduced below)?

(45) However an over use of bias, opinion, and suggestive language can rob readers of the chance to interpret the news events for themselves.

a. NO CHANGE
b. However, an over use of bias,
c. However, with too much bias,
d. However, an overuse of bias,

Questions 46–54 are based on the following passage:

(46) <u>One of the icon's of romantic and science fiction literature</u> remains Mary Shelley's classic, *Frankenstein, or The Modern Prometheus*. Schools throughout the world still teach the book in literature and philosophy courses. Scientific communities also engage in discussion on the novel. But why? Besides the novel's engaging (47) <u>writing style the story's central theme</u> remains highly relevant in a world of constant discovery and moral dilemmas. Central to the core narrative is the (48) <u>struggle between enlightenment and the cost of overusing power.</u>

The subtitle, *The Modern Prometheus*, encapsulates the inner theme of the story more than the main title of *Frankenstein*. As with many romantic writers, Shelley invokes the classical myths and (49) <u>symbolism of Ancient Greece and Rome to high light core ideas.</u> Looking deeper into the myth of Prometheus sheds light not only on the character of Frankenstein (50) <u>but also poses a psychological dilemma to the audience.</u> Prometheus is the titan who gave fire to mankind. (51) <u>However, more than just fire he gave people knowledge and power.</u> The power of fire advanced civilization. Yet, for giving fire to man, Prometheus is (52) <u>punished by the gods bound to a rock and tormented for his act</u>. This is clearly a parallel to Frankenstein—he is the modern Prometheus.

Frankenstein's quest for knowledge becomes an obsession. It leads him to literally create new life, breaking the bounds of conceivable science to illustrate that man can create life out of nothing. (53) <u>Yet he ultimately faltered as a creator,</u> abandoning his progeny in horror of what he created. Frankenstein then suffers his creature's wrath, (54) <u>the result of his pride, obsession for power and lack of responsibility.</u>

Shelley isn't condemning scientific achievement. Rather, her writing reflects that science and discovery are good things, but, like all power, it must be used wisely. The text alludes to the message that one must have reverence for nature and be mindful of the potential consequences. Frankenstein did not take responsibility or even consider how his actions would affect others. His scientific brilliance ultimately led to suffering.

Based on an excerpt from Frankenstein by Mary Shelley

46. Which of the following would be the best choice for this sentence (reproduced below)?

(46) <u>One of the icon's of romantic and science fiction literature</u> remains Mary Shelley's classic, *Frankenstein, or The Modern Prometheus.*

a. NO CHANGE
b. One of the icons of romantic and science fiction literature
c. One of the icon's of romantic, and science fiction literature,
d. The icon of romantic and science fiction literature

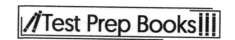

47. Which of the following would be the best choice for this sentence (reproduced below)?

Besides the novel's engaging (47) <u>writing style the story's central theme</u> remains highly relevant in a world of constant discovery and moral dilemmas.

a. NO CHANGE
b. writing style the central theme of the story
c. writing style, the story's central theme
d. the story's central theme's writing style

48. Which of the following would be the best choice for this sentence (reproduced below)?

Central to the core narrative is the (48) <u>struggle between enlightenment and the cost of overusing power.</u>

a. NO CHANGE
b. struggle between enlighten and the cost of overusing power.
c. struggle between enlightenment's cost of overusing power.
d. struggle between enlightening and the cost of overusing power.

49. Which of the following would be the best choice for this sentence (reproduced below)?

As with many romantic writers, Shelley invokes the classical myths and (49) <u>symbolism of Ancient Greece and Rome to high light core ideas.</u>

a. NO CHANGE
b. symbolism of Ancient Greece and Rome to highlight core ideas.
c. symbolism of ancient Greece and Rome to highlight core ideas.
d. symbolism of Ancient Greece and Rome highlighting core ideas.

50. Which of the following would be the best choice for this sentence (reproduced below)?

Looking deeper into the myth of Prometheus sheds light not only on the character of Frankenstein (50) <u>but also poses a psychological dilemma to the audience.</u>

a. NO CHANGE
b. but also poses a psychological dilemma with the audience.
c. but also poses a psychological dilemma for the audience.
d. but also poses a psychological dilemma there before the audience.

51. Which of the following would be the best choice for this sentence (reproduced below)?

(51) <u>However, more than just fire he gave people knowledge and power.</u>

a. NO CHANGE
b. However, more than just fire he gave people, knowledge, and power.
c. However, more than just fire, he gave people knowledge and power.
d. Besides actual fire, Prometheus gave people knowledge and power.

52. Which of the following would be the best choice for this sentence (reproduced below)?

Yet, for giving fire to man, Prometheus is (52) punished by the gods bound to a rock and tormented for his act.

a. NO CHANGE
b. punished by the gods, bound to a rock and tormented for his act.
c. bound to a rock and tormented as punishment by the gods.
d. punished for his act by being bound to a rock and tormented as punishment from the gods.

53. Which of the following would be the best choice for this sentence (reproduced below)?

(53) Yet he ultimately faltered as a creator, abandoning his progeny in horror of what he created.

a. NO CHANGE
b. Yet, he ultimately falters as a creator by
c. Yet, he ultimately faltered as a creator,
d. Yet he ultimately falters as a creator by

54. Which of the following would be the best choice for this sentence (reproduced below)?

Frankenstein then suffers his creature's wrath, (54) the result of his pride, obsession for power and lack of responsibility.

a. NO CHANGE
b. the result of his pride, obsession for power and lacking of responsibility.
c. the result of his pride, obsession for power, and lack of responsibility.
d. the result of his pride and also his obsession for power and lack of responsibility.

Questions 55–63 are based on the following passage:

The power of legends continues to enthrall our imagination, provoking us both to wonder and explore. (55) Who doesnt love a good legend? Some say legends never (56) die and this is certainly the case for the most legendary creature of all, Bigfoot. To this day, people still claim sightings of the illusive cryptid. Many think of Bigfoot as America's monster, yet many nations have legends of a similar creature. In my own research I have found that Australia has the Yowie, China has the Yerin, and Russia has the Almas. (57) Their all over the world, the bigfoots and the legends tied to them. Does this mean they could exist?

There are many things to consider when addressing (58) this question but the chief factor is whether there is credible evidence. (59) For science to formally recognize that such a species exists, there needs to be physical proof. While people have found supposed footprints and even (60) captured photos and film of the creature, this validity of such evidence is up for debate. There is room for uncertainty. Most visual evidence is out of focus, thus (61) there is often skepticism whether such images are real. Some researchers have even claimed to have hair and blood samples, but still there is doubt in the scientific community. The reason is simple: there needs to be a body or living specimen found and actively studied in order to prove the Bigfoots' existence.

Yet, one cannot ignore the fact that (62) <u>hundreds of witnesses continuing to describe a creature</u> with uniform features all over the world. These bigfoot sightings aren't a modern occurrence either. Ancient civilizations have reported (63) <u>seeing Bigfoot as well including Native Americans</u>. It is from Native Americans that we gained the popular term Sasquatch, which is the primary name for the North American bigfoot. How does their testimony factor in? If indigenous people saw these animals, could they not have existed at some point? After all, when Europeans first arrived in Africa, they disbelieved the native accounts of the gorilla. But sure enough, Europeans eventually found gorillas and collected a body.

55. Which of the following would be the best choice for this sentence (reproduced below)?

(55) <u>Who doesnt love a good legend?</u>

a. NO CHANGE
b. Who does not love a good legend?
c. A good legend, who doesn't love one?
d. Who doesn't love a good legend?

56. Which of the following would be the best choice for this sentence (reproduced below)?

Some say legends never (56) <u>die and this is certainly the case</u> for the most legendary creature of all, Bigfoot.

a. NO CHANGE
b. die, and this is certainly the case
c. die; this is certainly the case
d. die. This is certainly the case

57. Which of the following would be the best choice for this sentence (reproduced below)?

(57) <u>Their all over the world, the</u> bigfoots and the legends tied to them.

a. NO CHANGE
b. There all over the world, the
c. They're all over the world, the
d. All over the world they are, the

58. Which of the following would be the best choice for this sentence (reproduced below)?

There are many things to consider when addressing (58) <u>this question but the chief factor</u> is whether there is credible evidence.

a. NO CHANGE
b. this question, but the chief factor
c. this question however the chief factor
d. this question; but the chief factor

59. Which of the following would be the best choice for this sentence (reproduced below)?

(59) <u>For science to formally recognize that such a species exists, there needs to be physical proof.</u>

a. NO CHANGE
b. Physical proof are needed in order for science to formally recognize that such a species exists.
c. For science to formally recognize that such a species exists there needs to be physical proof.
d. For science, to formally recognize that such a species exists, there needs to be physical proof.

60. Which of the following would be the best choice for this sentence (reproduced below)?

While people have found supposed footprints and even (60) <u>captured photos and film of the creature, this validity of such evidence is up for debate.</u>

a. NO CHANGE
b. captured photos and film of the creature. This validity of such evidence is up for debate.
c. captured photos and film of the creature, the validities of such evidence is up for debate.
d. captured photos and film of the creature, the validity of such evidence is up for debate.

61. Which of the following would be the best choice for this sentence (reproduced below)?

Most visual evidence is out of focus, thus there is (61) <u>often skepticism whether such images are real.</u>

a. NO CHANGE
b. often skepticism whether such images are real.
c. there is often skepticism, whether such images are real.
d. there is often skepticism weather such images are real.

62. Which of the following would be the best choice for this sentence (reproduced below)?

Yet, one cannot ignore the fact that (62) <u>hundreds of witnesses continuing to describe a creature</u> with uniform features all over the world.

a. NO CHANGE
b. hundreds of witnesses continuing to describing a creature
c. hundreds of witnesses continue to describe a creature
d. hundreds of the witnesses continue to described a creature

63. Which of the following would be the best choice for this sentence (reproduced below)?

Ancient civilizations have reported (63) <u>seeing Bigfoot as well including Native Americans.</u>

a. NO CHANGE
b. seeing Bigfoot, Native Americans as well.
c. seeing Bigfoot also the Native Americans.
d. seeing Bigfoot, including Native Americans.

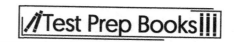

Questions 64–70 are based on the following passage:

I have to admit that when my father bought a recreational vehicle (RV), I thought he was making a huge mistake. I didn't really know anything about RVs, but I knew that my dad was as big a "city slicker" as there was. (64) In fact, I even thought he might have gone a little bit crazy. On trips to the beach, he preferred to swim at the pool, and whenever he went hiking, he avoided touching any plants for fear that they might be poison ivy. Why would this man, with an almost irrational fear of the outdoors, want a 40-foot camping behemoth?

(65) The RV was a great purchase for our family and brought us all closer together. Every morning (66) we would wake up, eat breakfast, and broke camp. We laughed at our own comical attempts to back The Beast into spaces that seemed impossibly small. (67) We rejoiced as "hackers." When things inevitably went wrong and we couldn't solve the problems on our own, we discovered the incredible helpfulness and friendliness of the RV community. (68) We even made some new friends in the process.

(69) Above all, it allowed us to share adventures. While traveling across America, which we could not have experienced in cars and hotels. Enjoying a campfire on a chilly summer evening with the mountains of Glacier National Park in the background or waking up early in the morning to see the sun rising over the distant spires of Arches National Park are memories that will always stay with me and our entire family. (70) Those are also memories that my siblings and me have now shared with our own children.

64. Which of the following would be the best choice for this sentence (reproduced below)?

(64) In fact, I even thought he might have gone a little bit crazy.

a. NO CHANGE
b. Move the sentence so that it comes before the preceding sentence.
c. Move the sentence to the end of the first paragraph.
d. Omit the sentence.

65. In context, which is the best version of the underlined portion of this sentence (reproduced below)?

(65) The RV was a great purchase for our family and brought us all closer together.

a. NO CHANGE
b. Not surprisingly, the RV
c. Furthermore, the RV
d. As it turns out, the RV

66. Which is the best version of the underlined portion of this sentence (reproduced below)?

Every morning (66) we would wake up, eat breakfast, and broke camp.

a. NO CHANGE
b. we would wake up, eat breakfast, and break camp.
c. would we wake up, eat breakfast, and break camp?
d. we are waking up, eating breakfast, and breaking camp.

67. Which is the best version of the underlined portion of this sentence (reproduced below)?

(67) <u>We rejoiced as "hackers."</u>

a. NO CHANGE
b. To a nagging problem of technology, we rejoiced as "hackers."
c. We rejoiced when we figured out how to "hack" a solution to a nagging technological problem.
d. To "hack" our way to a solution, we had to rejoice.

68. Which is the best version of the underlined portion of this sentence (reproduced below)?

(68) <u>We even made some new friends in the process.</u>

a. NO CHANGE
b. In the process was the friends we were making.
c. We are even making some new friends in the process.
d. We will make new friends in the process.

69. Which is the best version of the underlined portion of this sentence (reproduced below)?

(69) <u>Above all, it allowed us to share adventures. While traveling across America,</u> which we could not have experienced in cars and hotels.

a. NO CHANGE
b. Above all, it allowed us to share adventures while traveling across America
c. Above all, it allowed us to share adventures; while traveling across America
d. Above all, it allowed us to share adventures—while traveling across America

70. Which is the best version of the underlined portion of this sentence (reproduced below)?

(70) <u>Those are also memories that my siblings and me</u> have now shared with our own children.

a. NO CHANGE
b. Those are also memories that me and my siblings
c. Those are also memories that my siblings and I
d. Those are also memories that I and my siblings

Questions 71–75 are based on the following passage:

We live in a savage world; that's just a simple fact. It is a time of violence, when the need for self-defense is imperative. (71) <u>Martial arts, like Jiu-Jitsu, still play a vital role in ones survival.</u> (72) <u>Jiu-Jitsu, however doesn't justify kicking people around</u>, even when being harassed or attacked. Today, laws prohibit the (73) <u>use of unnecessary force in self-defense; these serve to eliminate</u> beating someone to a pulp once they have been neutralized. Such laws are needed. Apart from being unnecessary to continually strike a person when (74) <u>their down, its immoral</u>. Such over-aggressive retaliation turns the innocent into the aggressor. Jiu-Jitsu provides a way for defending oneself while maintaining the philosophy of restraint and self-discipline. (75) <u>Integrated into its core philosophy,</u> Jiu-Jitsu tempers the potential to do great physical harm with respect for that power and for life.

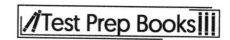

71. Which of the following would be the best choice for this sentence (reproduced below)?

(71) <u>Martial arts, like Jiu-Jitsu, still play a vital role in ones survival.</u>

a. NO CHANGE
b. Martial arts, like Jiu-Jitsu, still play a vital role in one's survival.
c. Martial arts, like Jiu-Jitsu still play a vital role in ones survival.
d. Martial arts, like Jiu-Jitsu, still plays a vital role in one's survival.

72. Which of the following would be the best choice for this sentence (reproduced below)?

(72) <u>Jiu-Jitsu, however doesn't justify kicking people around,</u> even when being harassed or attacked.

a. NO CHANGE
b. Jiu-Jitsu, however, isn't justified by kicking people around,
c. However, Jiu-Jitsu doesn't justify kicking people around,
d. Jiu-Jitsu however doesn't justify kicking people around,

73. Which of the following would be the best choice for this sentence (reproduced below)?

Today, laws prohibit the (73) <u>use of unnecessary force in self-defense; these serve to eliminate</u> beating someone to a pulp once they have been neutralized.

a. NO CHANGE
b. use of unnecessary force in self-defense serving to eliminate
c. use of unnecessary force, in self-defense, these serve to eliminate
d. use of unnecessary force. In self-defense, these serve to eliminate

74. Which of the following would be the best choice for this sentence (reproduced below)?

Apart from being unnecessary to continually strike a person when (74) <u>their down, its immoral.</u>

a. NO CHANGE
b. their down, it's immoral.
c. they're down, its immoral.
d. they're down, it's immoral.

75. Which of the following would be the best choice for this sentence (reproduced below)?

(75 <u>Integrated into its core philosophy,</u> Jiu-Jitsu tempers the potential to do great physical harm with respect for that power, and for life.

a. NO CHANGE
b. Integrated into its core philosophy,
c. Integrated into it's core philosophy,
d. Integrated into its' core philosophy,

Math

1. $3.4 + 2.35 + 4 =$
   a. 5.35
   b. 9.2
   c. 9.75
   d. 10.25
   e. 11.15

2. $5.88 \times 3.2 =$
   a. 18.816
   b. 16.44
   c. 20.352
   d. 17
   e. 19.25

3. $\frac{3}{25} =$
   a. 0.15
   b. 0.1
   c. 0.9
   d. 0.75
   e. 0.12

4. Which of the following is largest?
   a. 0.45
   b. 0.096
   c. 0.3
   d. 0.313
   e. 0.25

5. Which of the following is NOT a way to write 40 percent of N?
   a. $(0.4)N$

   b. $\frac{2}{5}N$

   c. $40N$

   d. $\frac{4N}{10}$

   e. $\frac{8N}{20}$

6. Which is closest to 17.8×9.9?
   a. 140
   b. 180
   c. 200
   d. 350
   e. 400

7. A student gets an 85% on a test with 20 questions. How many answers did the student solve correctly?

    a. 15

    b. 16

    c. 17

    d. 18

    e. 19

8. Four people split a bill. The first person pays for $\frac{1}{5}$, the second person pays for $\frac{1}{4}$, and the third person pays for $\frac{1}{3}$. What fraction of the bill does the fourth person pay?

    a. $\frac{13}{60}$

    b. $\frac{47}{60}$

    c. $\frac{1}{4}$

    d. $\frac{4}{15}$

    e. $\frac{1}{2}$

9. 6 is 30% of what number?

    a. 18

    b. 20

    c. 24

    d. 26

    e. 22

10. $3\frac{2}{3} - 1\frac{4}{5} =$

    a. $1\frac{13}{20}$

    b. $\frac{14}{15}$

    c. $2\frac{2}{3}$

    d. $\frac{4}{5}$

    e. $1\frac{13}{15}$

11. What is $\frac{420}{98}$ rounded to the nearest integer?

    a. 4

    b. 3

    c. 5

    d. 6

    e. 2

12. $4\frac{1}{3} + 3\frac{3}{4} =$

    a. $6\frac{5}{12}$

    b. $8\frac{1}{12}$

    c. $8\frac{2}{3}$

    d. $7\frac{7}{12}$

    e. $9\frac{1}{12}$

13. Five of six numbers have a sum of 25. The average of all six numbers is 6. What is the sixth number?
    a. 8
    b. 10
    c. 11
    d. 12
    e. 13

14. $52.3 \times 10^{-3} =$
    a. 0.00523
    b. 0.0523
    c. 0.523
    d. 523
    e. 5,230

15. If $\frac{5}{2} \div \frac{1}{3} = n$, then n is between:
    a. 5 and 7
    b. 1 and 3
    c. 9 and 11
    d. 3 and 5
    e. 7 and 9

16. A closet is filled with red, blue, and green shirts. If $\frac{1}{3}$ of the shirts are green and $\frac{2}{5}$ are red, what fraction of the shirts are blue?

    a. $\frac{4}{15}$

    b. $\frac{1}{5}$

    c. $\frac{7}{15}$

    d. $\frac{1}{2}$

    e. $\frac{1}{3}$

17. Shawna buys $2\frac{1}{2}$ gallons of paint. If she uses $\frac{1}{3}$ of it on the first day, how much does she have left?

   a. $1\frac{5}{6}$ gallons

   b. $1\frac{1}{2}$ gallons

   c. $1\frac{2}{3}$ gallons

   d. 2 gallons

   e. $2\frac{2}{3}$ gallons

18. If $6t + 4 = 16$, what is t?
   a. 1
   b. 2
   c. 3
   d. 4
   e. 5

19. There are $4x + 1$ treats in each party favor bag. If a total of $60x + 15$ treats is distributed, how many bags are given out?
   a. 15
   b. 16
   c. 20
   d. 22
   e. 24

20. In an office, there are 50 workers. A total of 60% of the workers are women, and the chances of a woman wearing a skirt is 50%. If no men wear skirts, how many workers are wearing skirts?
   a. 12
   b. 15
   c. 16
   d. 20
   e. 21

21. If $2x + 6 = 20$, what is x?
   a. 3
   b. 4
   c. 7
   d. 9
   e. 11

22. The variable y is directly proportional to x. If $y = 3$ when $x = 5$, then what is y when $x = 20$?
   a. 10
   b. 12
   c. 14
   d. 16
   e. 18

23. Apples cost $2 each, while oranges cost $3 each. Maria purchased 10 fruits in total and spent $22. How many apples did she buy?
    a. 5
    b. 6
    c. 7
    d. 4
    e. 8

24. What are the polynomial roots of $x^2 + x - 2$?
    a. 1 and -2
    b. -1 and 2
    c. 2 and -2
    d. 9 and 13
    e. 3 and 9

25. What is the *y*-intercept of $y = x^{5/3} + (x - 3)(x + 1)$?
    a. 3.5
    b. 7.6
    c. -3
    d. -15.1
    e. 3

26. Suppose $\frac{x+2}{x} = 2$. What is *x*?
    a. -1
    b. 0
    c. 2
    d. 4
    e. -2

27. A rectangle has a length that is 5 feet longer than three times its width. If the perimeter is 90 feet, what is the length in feet?
    a. 10
    b. 20
    c. 25
    d. 35
    e. 40

28. Five students take a test. The scores of the first four students are 80, 85, 75, and 60. If the median score is 80, which of the following could NOT be the score of the fifth student?
    a. 60
    b. 80
    c. 85
    d. 100
    e. 90

29. Change $3\frac{3}{5}$ to a decimal.
   a. 3.6
   b. 4.67
   c. 5.3
   d. 0.28
   e. 6.3

30. If a car can travel 300 miles in 4 hours, how far can it go in an hour and a half?
   a. 100 miles
   b. 112.5 miles
   c. 135.5 miles
   d. 150 miles
   e. 155 miles

31. Which measure for the center of a small sample set is most affected by outliers?
   a. Mean
   b. Median
   c. Mode
   d. Range
   e. None of these

32. Given the value of a given stock at monthly intervals, which graph should be used to best represent the trend of the stock?
   a. Box plot
   b. Line plot
   c. Scatter plot
   d. Circle graph
   e. Line graph

33. What is the probability of randomly picking the winner and runner-up from a race of 4 horses and distinguishing which is the winner?
   a. $\frac{1}{4}$
   b. $\frac{1}{2}$
   c. $\frac{1}{16}$
   d. $\frac{1}{12}$
   e. $\frac{1}{3}$

34. Which of the following could be used in the classroom to show $\frac{3}{7} < \frac{5}{6}$ is a true statement?
   a. A bar graph
   b. A number line
   c. An area model
   d. Base 10 blocks
   e. A line graph

35. Add $103,678 + 487$
    a. 103,191
    b. 103,550
    c. 104,265
    d. 104,165
    e. 105,143

36. Add $1.001 + 5.629$
    a. 4.472
    b. 4.628
    c. 5.630
    d. 6.245
    e. 6.630

37. Add $143.77 + 5.2$
    a. 138.57
    b. 148.97
    c. 138.97
    d. 148.57
    e. 149.087

38. What is the next number in the following series: $1, 3, 6, 10, 15, 21, \dots$?
    a. 26
    b. 27
    c. 28
    d. 29
    e. 30

39. Add and express in reduced form $\frac{14}{33} + \frac{10}{11}$.
    a. $\frac{2}{11}$
    b. $\frac{6}{11}$
    c. $\frac{4}{3}$
    d. $\frac{44}{33}$
    e. $\frac{3}{4}$

40. 32 is 25% of what number?
    a. 64
    b. 128
    c. 12.65
    d. 8
    e. 155

41. Subtract $112,076 - 1,243$.
   a. 109,398
   b. 113,319
   c. 113,833
   d. 110,833
   e. 111,485

42. Carey bought 184 pounds of fertilizer to use on her lawn. Each segment of her lawn required $11\frac{1}{2}$ pounds of fertilizer to do a sufficient job. If a student were asked to determine how many segments could be fertilized with the amount purchased, what operation would be necessary to solve this problem?
   a. Multiplication
   b. Division
   c. Addition
   d. Subtraction
   e. Exponents

43. Subtract $701.1 - 52.33$.
   a. 753.43
   b. 648.77
   c. 652.77
   d. 638.43
   e. 786.34

44. Which of the following expressions best exemplifies the additive and subtractive identity?
   a. $5 + 2 - 0 = 5 + 2 + 0$
   b. $6 + x = 6 - 6$
   c. $9 - 9 = 0$
   d. $8 + 2 = 10$
   e. $7 + 8 = 15$

45. Which of the following is an equivalent measurement for 1.3 cm?
   a. 0.13 m
   b. 13 m
   c. 0.13 mm
   d. 0.013 mm
   e. 0.013 m

46. Divide $1,015 \div 1.4$.
   a. 7,250
   b. 0.725
   c. 7.25
   d. 725
   e. 575

47. At the store, Jan spends $90 on apples and oranges. Apples cost $1 each and oranges cost $2 each. If Jan buys the same number of apples as oranges, how many oranges did she buy?
    a. 20
    b. 25
    c. 30
    d. 35
    e. 40

48. Multiply 12.4×0.2.
    a. 12.6
    b. 2.48
    c. 12.48
    d. 2.6
    e. 4.28

49. Multiply $1,987 \times 0.05$.
    a. 9.935
    b. 99.35
    c. 993.5
    d. 999.35
    e. 93.95

50. A clock reads 5:00 am. What is the measure of the angle formed by the two hands of that clock?
    a. 300 degrees
    b. 150 degrees
    c. 75 degrees
    d. 210 degrees
    e. 285 degrees

51. Divide, express with a remainder $1,202 \div 44$.
    a. $27\frac{2}{7}$

    b. $2\frac{7}{22}$

    c. $7\frac{2}{7}$

    d. $27\frac{7}{22}$

    e. $22\frac{7}{22}$

52. What is the volume of a box with rectangular sides 5 feet long, 6 feet wide, and 3 feet high?
    a. 60 cubic feet
    b. 75 cubic feet
    c. 80 cubic feet
    d. 14 cubic feet
    e. 90 cubic feet

53. Divide 702 ÷ 2.6.
    a. 27
    b. 207
    c. 2.7
    d. 270
    e. 2700

54. A train traveling 50 miles per hour takes a trip lasting 3 hours. If a map has a scale of 1 inch per 10 miles, how many inches apart are the train's starting point and ending point on the map?
    a. 14
    b. 12
    c. 13
    d. 15
    e. 16

55. A traveler takes an hour to drive to a museum, spends 3 hours and 30 minutes there, and takes half an hour to drive home. What percentage of his or her time was spent driving?
    a. 15%
    b. 30%
    c. 40%
    d. 60%
    e. 70%

56. A truck is carrying three cylindrical barrels. Their bases have a diameter of 2 feet, and they have a height of 3 feet. What is the total volume of the three barrels in cubic feet?
    a. 3π
    b. 9π
    c. 12π
    d. 15π
    e. 18π

57. Greg buys a $10 lunch with 5% sales tax. He leaves a $2 tip after his bill. How much money does he spend?
    a. $12.50
    b. $12
    c. $13
    d. $13.25
    e. $13.50

58. Add 5,089 + 10,323
    a. 15,402
    b. 15,412
    c. 5,234
    d. 15,234
    e. 16,001

59. A teacher is showing students how to evaluate $5 \times 6 + 4 \div 2 - 1$. Which operation should be completed first?
 a. Multiplication
 b. Addition
 c. Division
 d. Subtraction
 e. Exponent

60. What is the definition of a factor of the number 36?
 a. A number that can be divided by 36 and have no remainder
 b. A number that can be added to 36 with no remainder
 c. A prime number that is multiplied times 36
 d. A number that 36 can be divided by and have no remainder
 e. A number that 36 can be multiplied by and have no remainder

Reading

Fiction

Questions 1–10 are based upon the following passage:

"Did you ever come across a protégé of his—one Hyde?" He asked.

"Hyde?" repeated Lanyon. "No. Never heard of him. Since my time."

That was the amount of information that the lawyer carried back with him to the great, dark bed on which he tossed to and fro until the small hours of the morning began to grow large. It was a night of little ease to his toiling mind, toiling in mere darkness and besieged by questions.

Six o'clock struck on the bells of the church that was so conveniently near to Mr. Utterson's dwelling, and still he was digging at the problem. Hitherto it had touched him on the intellectual side alone; but now his imagination also was engaged, or rather enslaved; and as he lay and tossed in the gross darkness of the night in the curtained room, Mr. Enfield's tale went by before his mind in a scroll of lighted pictures. He would be aware of the great field of lamps in a nocturnal city; then of the figure of a man walking swiftly; then of a child running from the doctor's; and then these met, and that human Juggernaut trod the child down and passed on regardless of her screams. Or else he would see a room in a rich house, where his friend lay asleep, dreaming and smiling at his dreams; and then the door of that room would be opened, the curtains of the bed plucked apart, the sleeper recalled, and, lo! There would stand by his side a figure to whom power was given, and even at that dead hour he must rise and do its bidding. The figure in these two phrases haunted the lawyer all night; and if at anytime he dozed over, it was but to see it glide more stealthily through sleeping houses, or move the more swiftly, and still the more smoothly, even to dizziness, through wider labyrinths of lamplighted city, and at every street corner crush a child and leave her screaming. And still the figure had no face by which he might know it; even in his dreams it had no face, or one that baffled him and melted before his eyes; and thus there it was that there sprung up and grew apace in the lawyer's mind a singularly strong, almost an inordinate, curiosity to behold the features of the real Mr. Hyde. If he could but once set eyes on him, he thought the mystery would lighten and perhaps roll altogether away, as was the habit of mysterious things when well examined. He might see a reason for his friend's strange preference or bondage, and even for the startling clauses of the will. And at least it would be a face worth seeing: the face of a man who was without bowels of mercy: a face which had but to show itself to raise up, in the mind of the unimpressionable Enfield, a spirit of enduring hatred.

From that time forward, Mr. Utterson began to haunt the door in the by-street of shops. In the morning before office hours, at noon when business was plenty and time scarce, at night under the face of the full city moon, by all lights and at all hours of solitude or concourse, the lawyer was to be found on his chosen post.

"If he be Mr. Hyde," he had thought, "I should be Mr. Seek."

Excerpt from The Strange Case of Dr. Jekyll and Mr. Hyde by Robert Louis Stevenson

1. What is the purpose of the use of repetition in the following passage?

   It was a night of little ease to his toiling mind, toiling in mere darkness and besieged by questions.

   a. It serves as a demonstration of the mental state of Mr. Lanyon.
   b. It is reminiscent of the church bells that are mentioned in the story.
   c. It mimics Mr. Utterson's ambivalence.
   d. It emphasizes Mr. Utterson's anguish in failing to identify Hyde's whereabouts.

2. What is the setting of the story in this passage?
   a. In the city
   b. On the countryside
   c. In a jail
   d. In a mental health facility

3. What can one infer about the meaning of the word *Juggernaut* from the author's use of it in the passage?
   a. It is an apparition that appears at daybreak.
   b. It scares children.
   c. It is associated with space travel.
   d. Mr. Utterson finds it soothing.

4. What is the definition of the word *haunt* in the following passage?

> From that time forward, Mr. Utterson began to haunt the door in the by-street of shops. In the morning before office hours, at noon when business was plenty and time scarce, at night under the face of the full city moon, by all lights and at all hours of solitude or concourse, the lawyer was to be found on his chosen post.

   a. To levitate
   b. To constantly visit
   c. To terrorize
   d. To daunt

5. The phrase *labyrinths of lamplighted city* contains an example of what?
   a. Hyperbole
   b. Simile
   c. Juxtaposition
   d. Alliteration

6. What can one reasonably conclude from the final comment of this passage?

> "If he be Mr. Hyde," he had thought, "I should be Mr. Seek."

   a. The speaker is considering a name change.
   b. The speaker is experiencing an identity crisis.
   c. The speaker has mistakenly been looking for the wrong person.
   d. The speaker intends to continue to look for Hyde.

7. The author's attitude toward the main subject of this passage can be described as:
   a. Intrigue
   b. Elation
   c. Animosity
   d. Rigidity

8. According to the passage, what is Mr. Utterson struggling with as he tosses and turns in bed?
   a. A murderer who is stalking Mr. Utterson since he moved to the city.
   b. The mystery surrounding a dark figure and the terrible crimes he commits.
   c. The cases he is involved in as a detective.
   d. A chronic illness that is causing Mr. Utterson to hallucinate.

9. According to the passage, why did Mr. Utterson start to haunt the doors by the street shops?
   a. He was looking for a long, lost love who he kept dreaming about.
   b. He was recently homeless, and the street shops offered him food to eat when he was hungry.
   c. He was looking for the dark, mysterious figure who he had been obsessing over in his sleep.
   d. He was looking for a thief that would regularly steal out of stores.

10. What point of view is the passage written in?
   a. First person
   b. Second person
   c. Third person limited
   d. Third person omniscient

Social Science

Questions 11–20 are based on the following passages:

Passage I

Lethal force, or deadly force, is defined as the physical means to cause death or serious harm to another individual. The law holds that lethal force is only accepted when you or another person are in immediate and unavoidable danger of death or severe bodily harm. For example, a person could be beating a weaker person in such a way that they are suffering severe enough trauma that could result in death or serious harm. This would be an instance where lethal force would be acceptable and possibly the only way to save that person from irrevocable damage.

Another example of when to use lethal force would be when someone enters your home with a deadly weapon. The intruder's presence and possession of the weapon indicate mal-intent and the ability to inflict death or severe injury to you and your loved ones. Again, lethal force can be used in this situation. Lethal force can also be applied to prevent the harm of another individual. If a woman is being brutally assaulted and is unable to fend off an attacker, lethal force can be used to defend her as a last-ditch effort. If she is in immediate jeopardy of rape, harm, and/or death, lethal force could be the only response that could effectively deter the assailant.

The key to understanding the concept of lethal force is the term *last resort*. Deadly force cannot be taken back; it should be used only to prevent severe harm or death. The law does distinguish whether the means of one's self-defense is fully warranted, or if the individual goes out of control in the process. If you continually attack the assailant after they are rendered incapacitated, this would be causing unnecessary harm, and the law can bring charges against you. Likewise, if you kill an attacker unnecessarily after defending yourself, you can be charged with murder. This would move lethal force beyond necessary defense, making it no longer a last resort but rather a use of excessive force.

Passage II

Assault is the unlawful attempt of one person to apply apprehension on another individual by an imminent threat or by initiating offensive contact. Assaults can vary, encompassing physical strikes, threatening body language, and even provocative language. In the case of the latter, even if a hand has not been laid, it is still considered an assault because of its threatening nature.

Let's look at an example: A homeowner is angered because his neighbor blows fallen leaves into his freshly mowed lawn. Irate, the homeowner gestures a fist to his fellow neighbor and threatens to bash his head in for littering on his lawn. The homeowner's physical motions and verbal threat heralds a physical threat against the other neighbor. These factors classify the homeowner's reaction as an assault. If the angry neighbor hits the threatening homeowner in retaliation, that would constitute an assault as well because he physically hit the homeowner.

Assault also centers on the involvement of weapons in a conflict. If someone fires a gun at another person, this could be interpreted as an assault unless the shooter acted in self-defense. If an individual drew a gun or a knife on someone with the intent to harm them, that would be considered assault. However, it's also considered an assault if someone simply aimed a weapon, loaded or not, at another person in a threatening manner.

11. What is the purpose of the second passage?
    a. To inform the reader about what assault is and how it is committed
    b. To inform the reader about how assault is a minor example of lethal force
    c. To disprove the previous passage concerning lethal force
    d. The author is recounting an incident in which they were assaulted

12. Which of the following situations, according to the passages, would not constitute an illegal use of lethal force?
    a. A disgruntled cashier yells obscenities at a customer.
    b. A thief is seen running away with stolen cash.
    c. A man is attacked in an alley by another man with a knife.
    d. A woman punches another woman in a bar.

13. Given the information in the passages, which of the following must be true about assault?
    a. Assault charges are more severe than unnecessary use of force charges.
    b. There are various forms of assault.
    c. Smaller, weaker people cannot commit assaults.
    d. Assault is justified only as a last resort.

14. Which of the following, if true, would most seriously undermine the explanation proposed by the author of Passage I in the third paragraph?
    a. An instance of lethal force in self-defense is not absolutely absolved from blame. The law considers the necessary use of force at the time it is committed.
    b. An individual who uses lethal force under necessary defense is in direct compliance of the law under most circumstances.
    c. Lethal force in self-defense should be forgiven in all cases for the peace of mind of the primary victim.
    d. The use of lethal force is not evaluated on the intent of the user but rather the severity of the primary attack that warranted self-defense.

15. Based on the passages, what can be inferred about the relationship between assault and lethal force?
    a. An act of lethal force always leads to a type of assault.
    b. An assault will result in someone using lethal force.
    c. An assault with deadly intent can lead to an individual using lethal force to preserve their well-being.
    d. If someone uses self-defense in a conflict, it is called deadly force; if actions or threats are intended, it is called assault.

16. Which of the following best describes the way the passages are structured?
    a. Both passages open by defining a legal concept and then continue to describe situations that further explain the concept.
    b. Both passages begin with situations, introduce accepted definitions, and then cite legal ramifications.
    c. Passage I presents a long definition while the Passage II begins by showing an example of assault.
    d. Both cite specific legal doctrines, then proceed to explain the rulings.

17. What can be inferred about the role of intent in lethal force and assault?
    a. Intent is irrelevant. The law does not take intent into account.
    b. Intent is vital for determining the lawfulness of using lethal force.
    c. Intent is very important for determining both lethal force and assault; intent is examined in both parties and helps determine the severity of the issue.
    d. The intent of the assailant is the main focus for determining legal ramifications; it is used to determine if the defender was justified in using force to respond.

18. The author uses the example in the second paragraph of Passage II in order to do what?
    a. To demonstrate two different types of assault by showing how each specifically relates to the other
    b. To demonstrate a single example of two different types of assault, then adding in the third type of assault in the example's conclusion
    c. To prove that the definition of lethal force is altered when the victim in question is a homeowner and his property is threatened
    d. To suggest that verbal assault can be an exaggerated crime by the law and does not necessarily lead to physical violence

19. As it is used in the second passage, the word *apprehension* most nearly means:
    a. Pain
    b. Exhaustion
    c. Fear
    d. Honor

20. One of the main purposes of the last paragraph in the first passage is to state:
    a. How assault is different when used in the home versus when it is used out in public.
    b. A specific example of lethal force so that the audience will know what it looks like.
    c. Why police officers defend those who use lethal force but do not defend those who use assault.
    d. The concept of lethal force as a last resort and the point at which it can cross a line from defense to manslaughter.

Humanities

Questions 21–30 are based upon the following passage:

> My Good Friends,—When I first imparted to the committee of the projected Institute my particular wish that on one of the evenings of my readings here the main body of my audience should be composed of working men and their families, I was animated by two desires; first, by the wish to have the great pleasure of meeting you face to face at this Christmas time, and accompany you myself through one of my little Christmas books; and second, by the wish to have an opportunity of stating publicly in your presence, and

in the presence of the committee, my earnest hope that the Institute will, from the beginning, recognise one great principle—strong in reason and justice—which I believe to be essential to the very life of such an Institution. It is, that the working man shall, from the first unto the last, have a share in the management of an Institution which is designed for his benefit, and which calls itself by his name.

I have no fear here of being misunderstood—of being supposed to mean too much in this. If there ever was a time when any one class could of itself do much for its own good, and for the welfare of society—which I greatly doubt—that time is unquestionably past. It is in the fusion of different classes, without confusion; in the bringing together of employers and employed; in the creating of a better common understanding among those whose interests are identical, who depend upon each other, who are vitally essential to each other, and who never can be in unnatural antagonism without deplorable results, that one of the chief principles of a Mechanics' Institution should consist. In this world, a great deal of the bitterness among us arises from an imperfect understanding of one another. Erect in Birmingham a great Educational Institution, properly educational; educational of the feelings as well as of the reason; to which all orders of Birmingham men contribute; in which all orders of Birmingham men meet; wherein all orders of Birmingham men are faithfully represented—and you will erect a Temple of Concord here which will be a model edifice to the whole of England.

Contemplating as I do the existence of the Artisans' Committee, which not long ago considered the establishment of the Institute so sensibly, and supported it so heartily, I earnestly entreat the gentlemen—earnest I know in the good work, and who are now among us—by all means to avoid the great shortcoming of similar institutions; and in asking the working man for his confidence, to set him the great example and give him theirs in return. You will judge for yourselves if I promise too much for the working man, when I say that he will stand by such an enterprise with the utmost of his patience, his perseverance, sense, and support; that I am sure he will need no charitable aid or condescending patronage; but will readily and cheerfully pay for the advantages which it confers; that he will prepare himself in individual cases where he feels that the adverse circumstances around him have rendered it necessary; in a word, that he will feel his responsibility like an honest man, and will most honestly and manfully discharge it. I now proceed to the pleasant task to which I assure you I have looked forward for a long time.

From Charles Dickens' speech in Birmingham in England on December 30, 1853 on behalf of the Birmingham and Midland Institute.

21. Which word is most closely synonymous with the word *patronage* as it appears in the following statement?

...that I am sure he will need no charitable aid or condescending patronage

a. Auspices
b. Aberration
c. Acerbic
d. Adulation

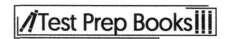

22. Which term is most closely aligned with the definition of the term *working man* as it is defined in the following passage?

> You will judge for yourselves if I promise too much for the working man, when I say that he will stand by such an enterprise with the utmost of his patience, his perseverance, sense, and support...

a. Plebeian
b. Viscount
c. Entrepreneur
d. Bourgeois

23. Which of the following statements most closely correlates with the definition of the term *working man* as it is defined in Question 22?
a. A working man is not someone who works for institutions or corporations, but someone who is well-versed in the workings of the soul.
b. A working man is someone who is probably not involved in social activities because the physical demand for work is too high.
c. A working man is someone who works for wages among the middle class.
d. The working man has historically taken to the field, to the factory, and now to the screen.

24. Based upon the contextual evidence provided in the passage above, what is the meaning of the term *enterprise* in the third paragraph?
a. Company
b. Courage
c. Game
d. Cause

25. The speaker addresses his audience as *My Good Friends.* What kind of credibility does this salutation give to the speaker?
a. The speaker is an employer addressing his employees, so the salutation is a way for the boss to bridge the gap between himself and his employees.
b. The speaker's salutation is one from an entertainer to his audience and uses the friendly language to connect to his audience before a serious speech.
c. The salutation is used ironically to give a somber tone to the serious speech that follows.
d. The speech is one from a politician to the public, so the salutation is used to grab the audience's attention.

26. According to the passage, what is the speaker's second desire for his time in front of the audience?
a. To read a Christmas story
b. For the working man to have a say in his institution, which is designed for his benefit
c. To have an opportunity to stand in their presence
d. For the life of the institution to be essential to the audience as a whole

27. The speaker's tone in the passage can be described as:
a. Happy and gullible
b. Lazy and entitled
c. Confident and informed
d. Angry and frustrated

28. One of the main purposes of the last paragraph is:
   a. To persuade the audience to support the Institute no matter what since it provided so much support to the working class.
   b. To market the speaker's new book while at the same time supporting the activities of the Institute.
   c. To inform the audience that the Institute is corrupt and will not help them out when the time comes to give them compensation.
   d. To provide credibility to the working man and share confidence in their ability to take on responsibilities if they are compensated appropriately.

29. According to the passage, what does the speaker wish to erect in Birmingham?
   a. An Educational Institution
   b. The Temple of Concord
   c. A Writing Workshop
   d. A VA Hospital

30. As it is used in the second paragraph, the word *antagonism* most nearly means:
   a. Conformity
   b. Opposition
   c. Affluence
   d. Scarcity

Natural Science

Questions 31–40 are based upon the following passage:

Three years ago, I think there were not many bird-lovers in the United States who believed it possible to prevent the total extinction of both egrets from our fauna. All the known rookeries accessible to plume-hunters had been totally destroyed. Two years ago, the secret discovery of several small, hidden colonies prompted William Dutcher, President of the National Association of Audubon Societies, and Mr. T. Gilbert Pearson, Secretary, to attempt the protection of those colonies. With a fund contributed for the purpose, wardens were hired and duly commissioned. As previously stated, one of those wardens was shot dead in cold blood by a plume hunter. The task of guarding swamp rookeries from the attacks of money-hungry desperadoes to whom the accursed plumes were worth their weight in gold, is a very chancy proceeding. There is now one warden in Florida who says that "before they get my rookery they will first have to get me."

Thus far the protective work of the Audubon Association has been successful. Now there are twenty colonies, which contain all told, about 5,000 egrets and about 120,000 herons and ibises which are guarded by the Audubon wardens. One of the most important is on Bird Island, a mile out in Orange Lake, central Florida, and it is ably defended by Oscar E. Baynard. To-day, the plume hunters who do not dare to raid the guarded rookeries are trying to study out the lines of flight of the birds, to and from their feeding-grounds, and shoot them in transit. Their motto is—"Anything to beat the law, and get the plumes." It is there that the state of Florida should take part in the war.

The success of this campaign is attested by the fact that last year a number of egrets were seen in eastern Massachusetts—for the first time in many years. And so to-day the question is, can the wardens continue to hold the plume-hunters at bay?

Excerpt from *Our Vanishing Wildlife* by William T. Hornaday

31. The author's use of first-person pronouns in the following text does NOT have which of the following effects?

   Three years ago, I think there were not many bird-lovers in the United States who believed it possible to prevent the total extinction of both egrets from our fauna.

   a. The phrase *I think* acts as a sort of hedging, where the author's tone is less direct and/or absolute.
   b. It allows the reader to more easily connect with the author.
   c. It encourages the reader to empathize with the egrets.
   d. It distances the reader from the text by overemphasizing the story.

32. What purpose does the quote serve at the end of the first paragraph?
   a. The quote shows proof of a hunter threatening one of the wardens.
   b. The quote lightens the mood by illustrating the colloquial language of the region.
   c. The quote provides an example of a warden protecting one of the colonies.
   d. The quote provides much needed comic relief in the form of a joke.

33. What is the meaning of the word *rookeries* in the following text?

   To-day, the plume hunters who do not dare to raid the guarded rookeries are trying to study out the lines of flight of the birds, to and from their feeding-grounds, and shoot them in transit.

   a. Houses in a slum area
   b. A place where hunters gather to trade tools
   c. A place where wardens go to trade stories
   d. A colony of breeding birds

34. What is on Bird Island?
   a. Hunters selling plumes
   b. An important bird colony
   c. Bird Island Battle between the hunters and the wardens
   d. An important egret with unique plumes

35. What is the main purpose of the passage?
   a. To persuade the audience to act in preservation of the bird colonies
   b. To show the effect hunting egrets has had on the environment
   c. To argue that the preservation of bird colonies has had a negative impact on the environment
   d. To demonstrate the success of the protective work of the Audubon Association

36. According to the passage, why are hunters trying to study the lines of flight of the birds?
   a. To study ornithology, one must know the lines of flight that birds take.
   b. To help wardens preserve the lives of the birds
   c. To have a better opportunity to hunt the birds
   d. To build their homes under the lines of flight because they believe it brings good luck

37. A year before the passage was written, where were a number of egrets seen?
    a. California
    b. Florida
    c. Eastern Massachusetts
    d. Western Texas

38. As it is used in the first paragraph, the word *commissioned* most nearly means:
    a. Appointed
    b. Compelled
    c. Beguiled
    d. Fortified

39. What happened two years before the passage was written?
    a. The plume hunters didn't dare to raid the rookeries, as they are heavily guarded.
    b. Twenty colonies have emerged as thousands of egrets are protected and make their homes in safe havens.
    c. The plume hunters tried to shoot the birds in their line of flight.
    d. Several hidden colonies were found which prompted Dutcher and Pearson to protect them.

40. As it is used in the second paragraph, the phrase *in transit* most nearly means:
    a. On a journey or trip
    b. To give authority to
    c. On the way to the destination
    d. To make angry

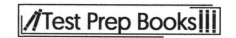

Science

Passage 1

Questions 1–5 pertain to Passage 1:

Scientists use the scientific method to investigate a theory or solve a problem. It includes four steps: observation, hypothesis, experiment, and conclusion. Observation occurs when the scientist uses one of their senses to identify what they want to study. A hypothesis is a conclusive sentence about what the scientist wants to research. It generally includes an explanation for the observations, can be tested experimentally, and predicts the outcome. The experiment includes the parameters for the testing that will occur. The conclusion will state whether or not the hypothesis was supported.

Scientist A would like to know how sunlight affects the growth of a plant. She says that more sunlight will cause the plant to grow faster. She sets up her experimental groups and tests her hypothesis over 11 days.

Figure 1 below shows the experimental data Scientist A collected over 11 days.

Figure 2 below represents the process of photosynthesis that occurs in plants.

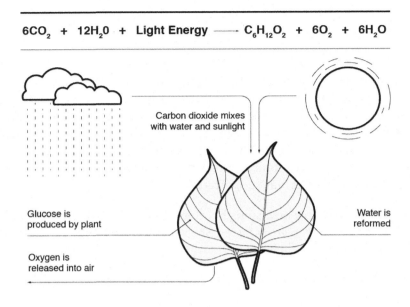

$$6CO_2 + 12H_2O + \text{Light Energy} \longrightarrow C_6H_{12}O_2 + 6O_2 + 6H_2O$$

Carbon dioxide mixes
with water and sunlight

Glucose is
produced by plant

Water is
reformed

Oxygen is
released into air

1. What is her hypothesis?
    a. More sunlight will cause the plant to grow faster.
    b. She will test her theory over 11 days.
    c. How sunlight affects plant growth.
    d. Plants do not grow well with one hour of sunlight per day.

2. How many experimental groups does she have?
    a. 1
    b. 3
    c. 4
    d. 11

3. What type of chart is represented in the first figure?
    a. Bar graph
    b. Line graph
    c. Pie chart
    d. Pictogram

4. What part of the photosynthesis reaction is provided directly by sunlight?
    a. Light energy
    b. H_2O
    c. CO_2
    d. Glucose

5. What should her conclusion be based on her experimental data?
    a. 5 hours of sunlight is optimal for plant growth.
    b. Plants should only be measured for 11 days.
    c. Less sunlight is better for plant growth.
    d. Providing plants with more sunlight makes them grow bigger.

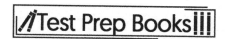

Passage 2

Questions 6–10 pertain to Passage 2:

The periodic table contains all known 118 chemical elements. The first 98 elements are found naturally while the remaining were synthesized by scientists. The elements are ordered according to the number of protons they contain, also known as their atomic number. For example, hydrogen has an atomic number of one and is found in the top left corner of the periodic table, whereas radon has an atomic number of 86 and is found closer on the right side of the periodic table, several rows down. The rows are called periods and the columns are called groups. The elements are arranged by similar chemical properties.

Each chemical element represents an individual atom. When atoms are linked together, they form molecules. The smallest molecule contains just two atoms, but molecules can also be very large and contain hundreds of atoms. In order to find the mass of a molecule, the atomic mass of each individual atom in the molecule must be added together.

Figure 1 below depicts the trends and commonalities between the elements that can be seen in the periodic table.

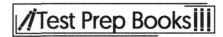

Figure 2 below shows what the information in each element's box represents.

Figure 3 below shows the periodic table with color coding according to the groups and periods.

6. What is the atomic mass of NaCl?
   a. 23
   b. 58.5
   c. 35.5
   d. 71

7. Which of the following elements is most electronegative?
   a. Ununoctium (Uuo)
   b. Francium (Fr)
   c. Hydrogen (H)
   d. Helium (He)

8. Which of the following elements has the lowest ionization energy?
   a. Potassium (K)
   b. Gallium (Ga)
   c. Sodium (Na)
   d. Aluminum (Al)

9. Which element has the fewest number of protons?
   a. Radon (Rn)
   b. Boron (B)
   c. Nitrogen (N)
   d. Hydrogen (H)

10. Scientist A needs a noble gas for her experiment. Which of these elements should she consider using?
   a. Nitrogen (N)
   b. Radon (Rn)
   c. Copper (Cu)
   d. Boron (B)

Passage 3

Questions 11–15 pertain to Passage 3:

Physical characteristics are controlled by genes. Each gene has two alleles, or variations. Generally, one allele is more dominant than the other allele and when one of each allele is present on the gene, the physical trait of the dominant allele will be expressed. The allele that is not expressed is called the recessive allele. Recessive alleles are expressed only when both alleles present on the gene are the recessive allele.

Punnett squares are diagrams that can predict the outcome of crossing different traits. In these diagrams, dominant alleles are represented by uppercase letters and recessive alleles are represented by lowercase letters.

Scientist A wants to grow white flowered plants and is doing a series of crossbreeding experiments. She had each plant genetically tested so she knows which alleles comprise each plant. The dominant flowers are red (A) and the recessive allele (a) produces white flowers.

Figure 1 below represents the different flowers that underwent crossbreeding during Round #1A

Round #1A

Crossbreeding #1A			Crossbreeding #2A			Crossbreeding #2A		
	A	**a**		**a**	**a**		**a**	**a**
A	AA	Aa	**A**	Aa	Aa	**A**	Aa	Aa
A	AA	Aa	**A**	Aa	Aa	**a**	aa	aa

Figure 2 below represents the number of flowers that were red and white after the first round of crossbreeding experiments.

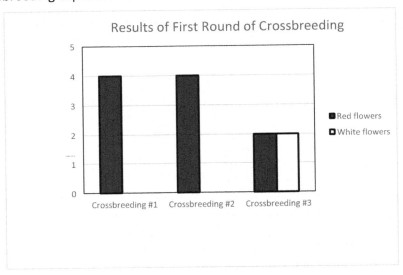

During her second round of crossbreeding, she adds in a plant of unknown genetic makeup with red flowers. She crosses it with a white-flowering plant. The results of her experiment are represented in the next figure.

Figure 3 represents the genetic results from the second round of crossbreeding.

Round #2		
	a	a
?	Aa	Aa
?	aa	aa

Scientist A takes offspring plants from Round #1A and crossbreeds them with each other and calls this Round #1B.

Figure 4 below represents the results of crossbreeding from Round #1B.

Crossbreeding #1B			Crossbreeding #2B			Crossbreeding #2B		
	A	a		A	a		a	a
A	AA	Aa	A	AA	Aa	a	aa	aa
A	AA	Aa	a	Aa	aa	a	aa	aa

11. Crossbreeding which two plants will give her the highest likelihood of obtaining some white plants right away in Round #1A?
  a. AA × Aa
  b. Aa × aa
  c. AA × aa
  d. AA × AA

12. What percentage of plants are white after the first crossbreeding reactions?
  a. 12
  b. 50
  c. 16.7
  d. 25

13. What is the genetic makeup of the unknown plant from the second round of crossbreeding?
  a. aa
  b. Aa
  c. AA
  d. Cannot be determined

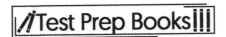

14. From which group of crossbreeding in Round #1B can she obtain 100% white flowers by the second generation?
   a. They are all equal.
   b. 1B
   c. 2B
   d. 3B

15. Which of her five senses did she use for the observation step of the scientific method here?
   a. Sight
   b. Smell
   c. Touch
   d. Hearing

Passage 4

Questions 16–20 pertain to Passage 4:

Rainforests cover approximately 6% of the Earth's surface. Tropical rainforests are found in five major areas of the world: Central America, South America, Central Africa, Asia stretching from India to islands in the Pacific Ocean, and Australia. All of these areas are warm and wet areas within ten degrees of the equator. They do not have a substantial dry season during the year.

Rainforests are large areas of jungle that get an abundance of rain. They comprise four layers, each with unique characteristics. The emergent layer is the highest layer and is made up of the tops of the tall trees. There is very good sunlight in this layer. The canopy layer is the next layer, just under the emergent layer. Here, there is some sun but not as much as the emergent layer. The next layer is the understory layer. This layer does not receive very much sunlight. The plants in this layer need to grow very large leaves to reach the sun. The bottom-most layer is the forest floor. Sunlight generally does not reach this layer, so plants do not grow here.

Figure 1 below represents the different layers of the rainforest.

Figure 2 below is a map of the rainforests on Earth and a map of Central America.

16. What essential part of photosynthesis, which is necessary for plant growth, is lacking on the forest floor and does not allow for plants to grow?
   a. Carbon
   b. Water
   c. Oxygen
   d. Sunlight

17. Which is a country in Central America that contains a rainforest?
   a. India
   b. Panama
   c. Madagascar
   d. Brazil

18. In which layer of the rainforest would birds fly around the most?
   a. Forest floor
   b. Understory layer
   c. Emergent layer
   d. Canopy layer

19. Giant taro plants have the largest leaves in the world that are approximately ten feet in length. In which layer of the rainforest do they reside?
   a. Emergent layer
   b. Canopy layer
   c. Forest floor
   d. Understory layer

20. Which of these plants, based on their listed requirements, would thrive in a rainforest climate?
   a. Giant water lily: warm temperatures, wet environment, has ability to grow large leaves
   b. Cactus: dry environment, long, hot season for growth
   c. Pine tree: dry and sandy soil, lots of sunlight
   d. Black-eyed susan: very hot temperatures, slightly moist soil

Passage 5

Questions 21–25 pertain to Passage 5:

An eclipse occurs when the light from one object in the solar system is completely or partially blocked by another object in the solar system. In 2017, the Earth was a part of two different types of eclipses. One was a solar eclipse, which occurs when the moon passes between the Sun and the Earth and blocks the Sun's light, making it dark during the daytime for several minutes. A total solar eclipse occurs when the moon completely covers the Sun. A partial solar eclipse occurs when the moon only covers part of the Sun. Solar eclipses should not be looked at directly because the Sun's rays can damage a person's eyes even though they appear to be dim while the eclipse is happening. Special viewing devices can be used to look at a solar eclipse indirectly, such as a pinhole camera facing away from the eclipse that allows light from the eclipse to pass through a hole in a piece of cardboard and its image to be reflected on a piece of white paper.

The other type of eclipse that the Earth was a part of was a lunar eclipse. This type of eclipse occurs when the moon passes behind the Earth, into its shadow. The moon is illuminated by the light of the sun, so when it is in the Earth's shadow, the moon becomes dim for a few hours during the night. This type of eclipse is safe to look at without any protection for your eyes.

Figure 1 below represents a solar eclipse.

Figure 2 below represents a lunar eclipse.

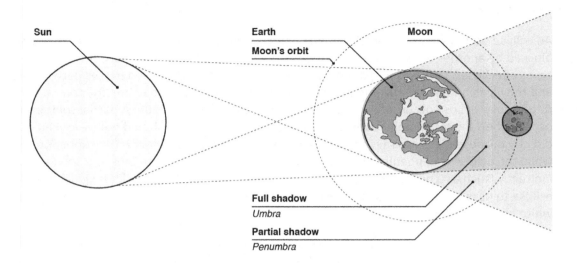

21. Which apparatus would be best to use to look at a solar eclipse?
    a. A telescope facing the eclipse
    b. A pinhole camera facing away from the eclipse
    c. Sunglasses facing the eclipse
    d. Binoculars facing the eclipse

22. What type of eclipse occurs when the moon comes between the Earth and Sun and covers the Sun's light completely?
    a. Total solar eclipse
    b. Partial lunar eclipse
    c. Total lunar eclipse
    d. Partial solar eclipse

23. What object in the solar system becomes dim during a lunar eclipse?
    a. Sun
    b. Earth
    c. Moon
    d. Earth and moon

24. Which type of eclipse could you observe directly using a telescope?
    a. Neither solar nor lunar
    b. Lunar only
    c. Both solar and lunar
    d. Solar only

25. Which type of eclipse is viewed during the daytime?
    a. Both solar and lunar
    b. Solar only
    c. Partial lunar
    d. Total lunar

Passage 6

Questions 26–30 pertain to Passage 6:

Phylogenetic trees are diagrams that map out the proposed evolutionary history of a species. They are branching diagrams that make it easy to see how scientists believe certain species developed from other species. The most recent proposed common ancestor between two species is the one before their lineages branch in the diagram. These diagrams do not attempt to include specific information about physical traits that were thought to be retained or disappeared during the evolutionary process.

Cladograms classify organisms based on their proposed common ancestry but are focused on their common physical traits. Branching points on these diagrams represent when a group of organisms is thought to have developed a new trait. Analogous features are those that have the same function but were not derived from a common ancestor. Homologous features have anatomical similarities, even if the function is no longer the same, due to a proposed common ancestor.

Figure 1 below is a phylogenetic tree of the Carnivora order.

Figure 2 below is a cladogram.

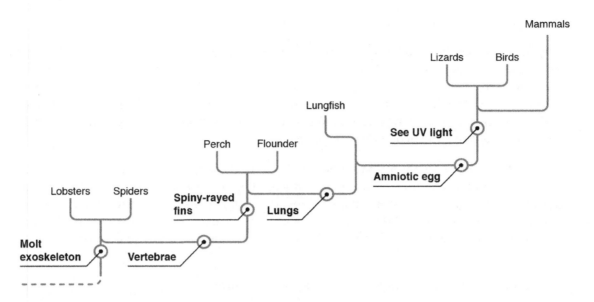

Figure 3 below shows a homologous feature between four different species with a proposed common ancestor.

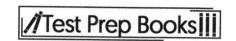

26. Which tool would you use to find out when a common ancestor of two species supposedly developed lungs?
    a. Phylogenetic tree
    b. Cladogram
    c. Punnett square
    d. Photographs

27. According to the earlier details, how are human arms and whale fins related?
    a. They are homologous structures from a common ancestor.
    b. They both have the same number of bones.
    c. They are analogous features.
    d. They are both covered in the same type of skin.

28. According to Figure 1, what common ancestry group do the striped skunk and European otter share?
    a. Mephitis
    b. Felidae
    c. Canidae
    d. Mustelidae

29. What trait do lizards and birds have in common according to Figure 2?
    a. Both see UV light
    b. Both have spiny-rayed fins
    c. Both molt an exoskeleton
    d. They do not have any traits in common.

30. According to Figure 1, at what level of organization are domestic cats and wolves related?
    a. Family
    b. Genus
    c. Order
    d. Species

Passage 7

Questions 31–35 pertain to Passage 7:

Scientists often use an assay called an enzyme-linked immunosorbent assay, or ELISA, to quantify specific substances within a larger sample. An ELISA works based on the specificity of an antibody to an antigen. One type of ELISA is called a sandwich ELISA. In this type of ELISA, a plate is coated with a capture antibody that adheres the antigen in the sample when it is added. Then the primary antibody is added and sticks to any antigen bound to the capture antibody. Next, a secondary antibody is added. Once it attaches to the primary antibody, it releases a colored tag that can be detected by a piece of laboratory equipment. If more color is released, it is indicative of more antigen having been present in the sample.

Figure 1 below describes how a sandwich ELISA works.

The cytokine protein IL-1β is a marker of inflammation in the body. Scientist A took samples from different locations within the body to find out where there was elevated inflammation in a patient.

Figure 2 below is a picture of the ELISA plate from Scientist A's experiment.

Figure 3 below is a graph of the results of Scientist A's experiment.

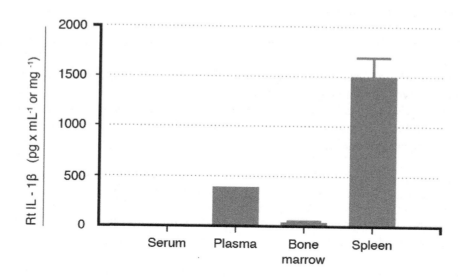

31. Which step of the ELISA allows for the color to be released for detection of the antigen?
    a. Addition of the antigen
    b. Addition of the primary antibody
    c. The presence of the capture antibody
    d. Addition of the secondary antibody

32. According to Figure 2, which row had the largest amount of antigen in the sample?
    a. Row 2
    b. Row 1
    c. Row 5
    d. Row 3

33. According to the ELISA results in Figure 3, which area of the body had the most inflammation?
    a. Serum
    b. Plasma
    c. Spleen
    d. Bone marrow

34. Which two antibodies sandwich the antigen in a sandwich ELISA?
    a. Capture antibody and secondary antibody
    b. Capture antibody and primary antibody
    c. Primary antibody and secondary antibody
    d. Two units of the secondary antibody

35. What is the purpose of an ELISA?
    a. Quantify specific substances within a larger sample
    b. Quantify all substances within a larger sample
    c. To create a colorful pattern with the samples
    d. To develop different antibodies

Passage 8

Questions 36–40 pertain to Passage 8:

Natural selection is the idea that certain traits make an individual have longer survival and higher reproduction rates than other individuals. It is based on the **phenotype**, or physical appearance, of the individual and not the **genotype**, or genetic makeup. There are three ways in which a phenotype can change due to natural selection. Directional selection occurs when one extreme of a phenotype is favored. Disruptive selection occurs when both extremes of a phenotype are favored. Stabilizing selection occurs when an intermediate phenotype is favored over either extreme phenotype.

Scenario 1: Mice live in an environment that has a mix of light and dark colored rocks. To avoid predators, the mice with intermediate color fur survive longer and produce more offspring.

Scenario 2: The Galapagos Islands experienced a drought and large, tough seeds became abundant. Finches developed large beaks to break up these seeds.

Scenario 3: In Cameroon, seeds are either large or small. Finches in Cameroon have either large beaks or small beaks. They are not found with medium-sized beaks.

36. What type of selection is described in Scenario 1?
    a. Stabilizing selection
    b. Directional selection
    c. Disruptive selection
    d. Color selection

37. What type of selection is described in Scenario 2?
    a. Beak-type
    b. Disruptive
    c. Stabilizing
    d. Directional

38. Why would it be hard for small-beaked finches in the Galapagos Island to survive after the drought?
    a. Too much sand got caught in the small beaks
    b. Beaks would not be able to break up large seeds
    c. Large-beaked finches would attack them
    d. Two extreme phenotypes can never be selected by natural selection

39. What type of selection is described in Scenario 3?
    a. Stabilizing
    b. Directional
    c. Disruptive
    d. Beak-type

40. Which statement is true about natural selection?
    a. Individuals are selected based on their genotype.
    b. An extreme phenotype is always selected.
    c. It only occurs after a drought.
    d. Individuals are selected based on phenotypes that are advantageous for survival and reproduction.

Writing

Please read the prompt below and answer in an essay format.

Some people feel that sharing their lives on social media sites such as Facebook, Instagram, and Snapchat is fine. They share every aspect of their lives, including pictures of themselves and their families, what they ate for lunch, who they are dating, and when they are going on vacation. They even say that if it's not on social media, it didn't happen. Other people believe that sharing so much personal information is an invasion of privacy and could prove dangerous. They think sharing personal pictures and details invites predators, cyberbullying, and identity theft.

1. Analyze and evaluate the passage given.

2. State and develop your own perspective.

3. Explain the relationship between your perspective and the one given.

Answer Explanations for Practice Test #1

English

1. C: Choice *C* correctly uses *from* to describe the fact that dogs are related to wolves. The word *through* is incorrectly used here, so Choice *A* is incorrect. Choice *B* makes no sense. Choice *D* unnecessarily changes the verb tense in addition to incorrectly using *through*.

2. B: Choice *B* is correct because the Oxford comma is applied, clearly separating the specific terms. Choice *A* lacks this clarity. Choice *C* is correct but too wordy since commas can be easily applied. Choice *D* doesn't flow with the sentence's structure.

3. D: Choice *D* correctly uses the question mark, fixing the sentence's main issue. Thus, Choice *A* is incorrect because questions do not end with periods. Choice *B*, although correctly written, changes the meaning of the original sentence. Choice *C* is incorrect because it completely changes the direction of the sentence, disrupts the flow of the paragraph, and lacks the crucial question mark.

4. A: Choice *A* is correct since there are no errors in the sentence. Choices *B* and *C* both have extraneous commas, disrupting the flow of the sentence. Choice *D* unnecessarily rearranges the sentence.

5. D: Choice *D* is correct because the commas serve to distinguish that *artificial selection* is just another term for *selective breeding* before the sentence continues. The structure is preserved, and the sentence can flow with more clarity. Choice *A* is incorrect because the sentence needs commas to avoid being a run-on. Choice *B* is close but still lacks the required comma after *selection*, so this is incorrect. Choice *C* is incorrect because the comma to set off the aside should be placed after *breeding* instead of *called*.

6. B: Choice *B* is correct because the sentence is talking about a continuing process. Therefore, the best modification is to add the word *to* in front of *increase*. Choice *A* is incorrect because this modifier is missing. Choice *C* is incorrect because, with the additional comma, the present tense of *increase* is inappropriate. Choice *D* makes more sense, but the tense is still not the best to use.

7. A: The sentence has no errors, so Choice *A* is correct. Choice *B* is incorrect because it adds an unnecessary comma. Choice *C* is incorrect because *advantage* should not be plural in this sentence without the removal of the singular *an*. Choice *D* is very tempting. While this would make the sentence more concise, this would ultimately alter the context of the sentence, which would be incorrect.

8. C: Choice *C* correctly uses *on to*, describing the way genes are passed generationally. The use of *into* is inappropriate for this context, which makes Choice *A* incorrect. Choice *B* is close, but *onto* refers to something being placed on a surface. Choice *D* doesn't make logical sense.

9. D: Choice *D* is correct, since only proper names should be capitalized. Because the name of a dog breed is not a proper name, Choice *A* is incorrect. In terms of punctuation, only one comma after *example* is needed, so Choices *B* and *C* are incorrect.

10. D: Choice *D* is the correct answer because "rather" acts as an interrupting word here and thus should be separated by commas. Choices *B* and *C* use commas unwisely, breaking the flow of the sentence.

11. B: Since the sentence can stand on its own without *Usually*, separating it from the rest of the sentence with a comma is correct. Choice *A* needs the comma after *Usually*, while Choice *C* uses commas incorrectly. Choice *D* is tempting, but changing *turn* to past tense goes against the rest of the paragraph.

12. A: In Choice *A*, the dependent clause *Sometimes in particularly dull seminars* is seamlessly attached with a single comma after *seminars*. Choice *B* contains too many commas. Choice *C* does not correctly combine the dependent clause with the independent clause. Choice *D* introduces too many unnecessary commas.

13. D: Choice *D* rearranges the sentence to be more direct and straightforward, so it is correct. Choice *A* needs a comma after *on*. Choice *B* introduces unnecessary commas. Choice *C* creates an incomplete sentence, since *Because I wasn't invested in what was going on* is a dependent clause.

14. C: Choice *C* is fluid and direct, making it the best revision. Choice *A* is incorrect because the construction is awkward and lacks parallel structure. Choice *B* is incorrect because of the unnecessary comma and period. Choice *D* is close, but its sequence is still awkward and overly complicated.

15. B: Choice *B* correctly adds a comma after *person* and cuts out the extraneous writing, making the sentence more streamlined. Choice *A* is poorly constructed, lacking proper grammar to connect the sections of the sentence correctly. Choice *C* inserts an unnecessary semicolon and doesn't enable this section to flow well with the rest of the sentence. Choice *D* is better but still unnecessarily long.

16. D: This sentence, though short, is a complete sentence. The only thing the sentence needs is an em-dash after "Easy." In this sentence the em-dash works to add emphasis to the word "Easy" and also acts in place of a colon, but in a less formal way. Therefore, Choice *D* is correct. Choices *A* and *B* lack the crucial comma, while Choice *C* unnecessarily breaks the sentence apart.

17. C: Choice *C* successfully fixes the construction of the sentence, changing *drawing* into *to draw*. Keeping the original sentence disrupts the flow, so Choice *A* is incorrect. Choice *B*'s use of *which* offsets the whole sentence. Choice *D* is incorrect because it unnecessarily expands the sentence content and makes it more confusing.

18. B: Choice *B* fixes the homophone issue. Because the author is talking about people, *their* must be used instead of *there*. This revision also appropriately uses the Oxford comma, separating and distinguishing *lives, world, and future*. Choice *A* uses the wrong homophone and is missing commas. Choice *C* neglects to fix these problems and unnecessarily changes the tense of *applies*. Choice *D* fixes the homophone but fails to properly separate *world* and *future*.

19. C: Choice *C* is correct because it fixes the core issue with this sentence: the singular *has* should not describe the plural *scientists*. Thus, Choice *A* is incorrect. Choices *B* and *D* add unnecessary commas.

20. D: Choice *D* correctly conveys the writer's intention of asking if, or *whether*, early perceptions of dinosaurs are still influencing people. Choice *A* makes no sense as worded. Choice *B* is better, but *how* doesn't coincide with the context. Choice *C* adds unnecessary commas.

21. A: Choice *A* is correct, as the sentence does not require modification. Choices *B* and *C* implement extra punctuation unnecessarily, disrupting the flow of the sentence. Choice *D* incorrectly adds a comma in an awkward location.

22. B: Choice *B* is the strongest revision, as adding *to explore* is very effective in both shortening the sentence and maintaining, even enhancing, the point of the writer. To explore is to seek understanding in order to gain knowledge and insight, which coincides with the focus of the overall sentence. Choice *A* is not technically incorrect, but it is overcomplicated. Choice *C* is a decent revision, but the sentence could still be more condensed and sharpened. Choice *D* fails to make the sentence more concise and inserts unnecessary commas.

23. D: Choice *D* correctly applies a semicolon to introduce a new line of thought while remaining in a single sentence. The comma after *however* is also appropriately placed. Choice *A* is a run-on sentence. Choice *B* is also incorrect because the single comma is not enough to fix the sentence. Choice *C* adds commas around *uncertain* which are unnecessary.

24. B: Choice *B* not only fixes the homophone issue from *its*, which is possessive, to *it's*, which is a contraction of *it is*, but also streamlines the sentence by adding a comma and eliminating *and*. Choice *A* is incorrect because of these errors. Choices *C* and *D* only fix the homophone issue.

25. A: Choice *A* is correct, as the sentence is fine the way it is. Choices *B* and *C* add unnecessary commas, while Choice *D* uses the possessive *its* instead of the contraction *it's*.

26. C: Choice *C* is correct because the phrase *even likely* is flanked by commas, creating a kind of aside, which allows the reader to see this separate thought while acknowledging it as part of the overall sentence and subject at hand. Choice *A* is incorrect because it seems to ramble after *even* due to a missing comma after *likely*. Choice *B* is better but inserting a comma after *that* warps the flow of the writing. Choice *D* is incorrect because there must be a comma after *plausible*.

27. D: Choice *D* strengthens the overall sentence structure while condensing the words. This makes the subject of the sentence, and the emphasis of the writer, much clearer to the reader. Thus, while Choice *A* is technically correct, the language is choppy and over-complicated. Choice *B* is better but lacks the reference to a specific image of dinosaurs. Choice *C* introduces unnecessary commas.

28. B: Choice *B* correctly joins the two independent clauses. Choice *A* is decent, but "that would be" is too verbose for the sentence. Choice *C* incorrectly changes the semicolon to a comma. Choice *D* splits the clauses effectively but is not concise enough.

29. A: Choice *A* is correct, as the original sentence has no error. Choices *B* and *C* employ unnecessary semicolons and commas. Choice *D* would be an ideal revision, but it lacks the comma after *Ransom* that would enable the sentence structure to flow.

30. D: By reorganizing the sentence, the context becomes clearer with Choice *D*. Choice *A* has an awkward sentence structure. Choice *B* offers a revision that doesn't correspond well with the original sentence's intent. Choice *C* cuts out too much of the original content, losing the full meaning.

31. C: Choice *C* fixes the disagreement between the singular *this* and the plural *viewpoints*. Choice *A*, therefore, is incorrect. Choice *B* introduces an unnecessary comma. In Choice *D*, *those* agrees with *viewpoints*, but neither agrees with *distinguishes*.

32. A: Choice *A* is direct and clear, without any punctuation errors. Choice *B* is well-written but too wordy. Choice *C* adds an unnecessary comma. Choice *D* is also well-written but much less concise than Choice *A*.

33. D: Choice *D* rearranges the sentence to improve clarity and impact, with *tempting* directly describing *idea*. On its own, Choice *A* is a run-on. Choice *B* is better because it separates the clauses, but it keeps an unnecessary comma. Choice *C* is also an improvement but still a run-on.

34. B: Choice *B* is the best answer simply because the sentence makes it clear that Un-man takes over and possesses Weston. In Choice *A*, these events sounded like two different things, instead of an action and result. Choices *C* and *D* make this relationship clearer, but the revisions don't flow very well grammatically.

35. D: Changing the phrase *after this* to *then* makes the sentence less complicated and captures the writer's intent, making Choice *D* correct. Choice *A* is awkwardly constructed. Choices *B* and *C* misuse their commas and do not adequately improve the clarity.

36. B: By starting a new sentence, the run-on issue is eliminated, and a new line of reasoning can be seamlessly introduced, making Choice *B* correct. Choice *A* is thus incorrect. While Choice *C* fixes the run-on via a semicolon, a comma is still needed after *this*. Choice *D* contains a comma splice. The independent clauses must be separated by more than just a comma, even with the rearrangement of the second half of the sentence.

37. C: Choice *C* condenses the original sentence while being more active in communicating the emphasis on changing times/media that the author is going for, so it is correct. Choice *A* is clunky because it lacks a comma after *today* to successfully transition into the second half of the sentence. Choice *B* inserts unnecessary commas. Choice *D* is a good revision of the underlined section, but not only does it not fully capture the original meaning, it also does not flow into the rest of the sentence.

38. B: Choice *B* clearly illustrates the author's point, with a well-placed semicolon that breaks the sentence into clearer, more readable sections. Choice *A* lacks punctuation. Choice *C* is incorrect because the period inserted after *question* forms an incomplete sentence. Choice *D* is a very good revision but does not make the author's point clearer than the original.

39. A: Choice *A* is correct: while the sentence seems long, it actually doesn't require any commas. The conjunction "that" successfully combines the two parts of the sentence without the need for additional punctuation. Choices *B* and *C* insert commas unnecessarily, incorrectly breaking up the flow of the sentence. Choice *D* alters the meaning of the original text by creating a new sentence, which is only a fragment.

40. C: Choice *C* correctly replaces *for* with *to*, the correct preposition for the selected area. Choice *A* is not the answer because of this incorrect preposition. Choice *B* is unnecessarily long and disrupts the original sentence structure. Choice *D* is also too wordy and lacks parallel structure.

41. D: Choice *D* is the answer because it inserts the correct punctuation to fix the sentence, linking the dependent and independent clauses. Choice *A* is therefore incorrect. Choice *B* is also incorrect since this revision only adds content to the sentence while lacking grammatical precision. Choice *C* overdoes the punctuation; only a comma is needed, not a semicolon.

42. B: Choice *B* correctly separates the section into two sentences and changes the word order to make the second part clearer. Choice *A* is incorrect because it is a run-on. Choice *C* adds an extraneous comma, while Choice *D* makes the run-on worse and does not coincide with the overall structure of the sentence.

43. C: Choice *C* is the best answer because of how the commas are used to flank *in earnest*. This distinguishes the side thought (*in earnest*) from the rest of the sentence. Choice *A* needs punctuation. Choice *B* inserts a semicolon in a spot that doesn't make sense, resulting in a fragmented sentence and lost meaning. Choice *D* is unnecessarily repetitive and creates a run-on.

44. A: Choice *A* is correct because the sentence contains no errors. The comma after *bias* successfully links the two halves of the sentence, and the use of *it's* is correct as a contraction of *it is*. Choice *B* creates a sentence fragment, while Choice *C* creates a run-on. Choice *D* incorrectly changes *it's* to *its*.

45. D: Choice *D* correctly inserts a comma after *However* and fixes *over use* to *overuse*—in this usage, it is one word. Choice *A* is therefore incorrect, as is Choice *B*. Choice *C* is a good revision but does not fit well with the rest of the sentence.

46. B: Choice *B* is correct because it removes the apostrophe from *icon's*, since the noun *icon* is not possessing anything. This conveys the author's intent of setting *Frankenstein* apart from other icons of the romantic and science fiction genres. Choices *A* and *C* are therefore incorrect. Choice *D* is a good revision but alters the meaning of the sentence—*Frankenstein* is one of the icons, not the sole icon.

47 C: Choice *C* correctly adds a comma after *style*, successfully joining the dependent and the independent clauses as a single sentence. Choice *A* is incorrect because the dependent and independent clauses remain unsuccessfully combined without the comma. Choices *B* and *D* do nothing to fix this.

48. A: Choice *A* is correct, as the sentence doesn't require changes. Choice *B* incorrectly changes the noun *enlightenment* into the verb *enlighten*. Choices *C* and *D* alter the original meaning of the sentence.

49. B: Choice *B* is correct, fixing the incorrect split of *highlight*. This is a polyseme, a word combined from two unrelated words to make a new word. On their own, *high* and *light* make no sense for the sentence, making Choice *A* incorrect. Choice *C* incorrectly decapitalizes *Ancient*—since it modifies *Greece* and works with the noun to describe a civilization, *Ancient Greece* functions as a proper noun, which should be capitalized. Choice *D* uses *highlighting*, a gerund, but the present tense of *highlight* is what works with the rest of the sentence; to make this change, a comma would be needed after *Rome*.

50. A: Choice *A* is correct, as *not only* and *but also* are correlative pairs. In this sentence, *but* successfully transitions the first part into the second half, making punctuation unnecessary. Additionally, the use of *to* indicates that an idea or challenge is being presented to the reader. Choice *B*'s *with*, *C*'s *for*, and *D*'s *there before* are not as active, meaning these revisions weaken the sentence.

51. D: Choice *D* is correct, adding finer details to help the reader understand exactly what Prometheus did and his impact: fire came with knowledge and power. Choice *A* lacks a comma after *fire*. Choice *B* inserts unnecessary commas since *people* is not part of the list *knowledge and power*. Choice *C* is a strong revision but could be confusing, hinting that the fire was knowledge and power itself, as opposed to being symbolized by the fire.

52. C: Choice *C* reverses the order of the section, making the sentence more direct. Choice *A* lacks a comma after *gods*, and although Choice *B* adds this, the structure is too different from the first half of the sentence to flow correctly. Choice *D* is overly complicated and repetitious in its structure even though it doesn't need any punctuation.

53. B: Choice *B* fixes the two problems of the sentence, changing *faltered* to present tense in agreement with the rest of the passage, and correctly linking the two dependent clauses. Choice *A* is therefore incorrect. Choice *C* does not correct the past tense of *faltered*. Choice *D* correctly adds the conjunction *by*, but it lacks a comma after the conjunction *Yet*.

54. C: Choice *C* successfully applies a comma after *power*, distinguishing the causes of Frankenstein's suffering and maintaining parallel structure. Choice *A* is thus incorrect. Choice *B* lacks the necessary punctuation and unnecessarily changes *lack* to a gerund. Choice *D* adds unnecessary wording, making the sentence more cumbersome.

55. D: Choice *D* correctly inserts an apostrophe into the contraction *doesn't*. Choice *A* is incorrect because of this omission. Choices *B* and *C* are better than the original but do not fit well with the informal tone of the passage.

56. B: Choice *B* is correct, successfully combining the two independent clauses of this compound sentence by adding a comma before "and" to create the effective pause and transition between clauses. Choice *A* does not join the independent clauses correctly. Choices *C* and *D* offer alternate ways of joining these clauses, but since "and" is already part of the sentence, adding the comma is the most logical choice. This also keeps the informal tone set by the rest of the passage.

57. C: Choice *C* correctly fixes the homophone issue of *their* and *they're*. *Their* implies ownership, which is not needed here. The author intends *they're*, a contraction of *they are*. Thus, Choice *A* is incorrect, as is Choice *B*, using the homophone *there*. Choice *D* eliminates the homophone issue altogether, but the sentence becomes more clunky because of that.

58. B: Choice *B* correctly joins the two independent clauses with a comma before *but*. Choice *A* is incorrect because, without the comma, it is a run-on sentence. Choice *C* also lacks punctuation and uses *however*, which should be reserved for starting a new sentence or perhaps after a semicolon. Choice *D* is incorrect because the semicolon throws off the sentence structure and is incorrectly used; the correct revision would have also removed *but*.

59. A: Choice *A* is correct because the sentence does not require modification. Choice *B* is incorrect because it uses the faulty subject/verb agreement, "Physical proof are." Choice *C* is incorrect because a comma would need to follow *exists*. Choice *D* is incorrect because the comma after *science* is unnecessary.

60. D: Choice *D* correctly changes *this* to *the* and retains *validity*, making it the right choice. Choices *A* and *B* keep *this*, which is not as specific as "the." Choice *C* incorrectly pluralizes *validity*.

61. A: Choice *A* is correct because the sentence is fine without revisions. Choice *B* is incorrect, since removing *there is* is unnecessary and confusing. Choice *C* is incorrect since it inserts an unnecessary comma. Choice *D* introduces a homophone issue: *weather* refers to climatic states and atmospheric events, while *whether* expresses doubt, which is the author's intent.

62. C: Choice *C* correctly changes *continuing* to the present tense. Choice *A* is incorrect because of this out-of-place gerund use. Choice *B* not only does not fix this issue but also incorrectly changes *describe* into a gerund. While Choice *D* correctly uses *continue*, *describe* is incorrectly put in the past tense.

63. D: Choice *D* is correct, since it eliminates the unnecessary *as well* and adds a comma to separate the given example, making the sentence more direct. Choice *A* seems repetitive with *as well*, since it has *including*, and at the least needs punctuation. Choice *B* is poorly constructed, taking out the clearer *including*. Choice *C* also makes little sense.

64. B: Move the sentence so that it comes before the preceding sentence. For this question, place the underlined sentence in each prospective choice's position. To keep it as-is is incorrect because the father "going crazy" doesn't logically follow the fact that he was a "city slicker." Choice *C* is incorrect because the sentence in question is not a concluding sentence and does not transition smoothly into the second paragraph. Choice *D* is incorrect because the sentence doesn't necessarily need to be omitted since it logically follows the very first sentence in the passage.

65. D: Choice *D* is correct because "As it turns out" indicates a contrast from the previous sentiment, that the RV was a great purchase. Choice *A* is incorrect because the sentence needs an effective transition from the paragraph before. Choice *B* is incorrect because the text indicates it *is* surprising that the RV was a great purchase because the author was skeptical beforehand. Choice *C* is incorrect because the transition "Furthermore" does not indicate a contrast.

66. B: This sentence calls for parallel structure. Choice *B* is correct because the verbs "wake," "eat," and "break" are consistent in tense and parts of speech. Choice *A* is incorrect because the words "wake" and "eat" are present tense while the word "broke" is in past tense. Choice *C* is incorrect because this turns the sentence into a question, which doesn't make sense within the context. Choice *D* is incorrect because it breaks tense with the rest of the passage. "Waking," "eating," and "breaking" are all present participles, and the context around the sentence is in past tense.

67. C: Choice *C* is correct because it is clear and fits within the context of the passage. Choice *A* is incorrect because "We rejoiced as 'hackers'" does not give a reason why hacking was rejoiced. Choice *B* is incorrect because it does not mention a solution being found and is therefore not specific enough. Choice *D* is incorrect because the meaning is eschewed by the helping verb "had to rejoice," and the sentence suggests that rejoicing was necessary to "hack" a solution.

68. A: The original sentence is correct because the verb tense as well as the meaning aligns with the rest of the passage. Choice *B* is incorrect because the order of the words makes the sentence more confusing than it otherwise would be. Choice *C* is incorrect because "We are even making" is in present tense. Choice *D* is incorrect because "We will make" is future tense. The surrounding text of the sentence is in past tense.

69. B: Choice *B* is correct because there is no punctuation needed if a dependent clause ("while traveling across America") is located behind the independent clause ("it allowed us to share adventures"). Choice *A* is incorrect because there are two dependent clauses connected and no independent clause, and a complete sentence requires at least one independent clause. Choice *C* is incorrect because of the same reason as Choice *A*. Semicolons have the same function as periods: there must be an independent clause on either side of the semicolon. Choice *D* is incorrect because the dash simply interrupts the complete sentence.

70. C: The rules for "me" and "I" is that one should use "I" when it is the subject pronoun of a sentence, and "me" when it is the object pronoun of the sentence. Break the sentence up to see if "I" or "me" should be used. To say "Those are memories that I have now shared" is correct, rather than "Those are memories that me have now shared." Choice *D* is incorrect because "my siblings" should come before "I."

71. B: Choice *B* is correct because it adds an apostrophe to *ones*, which indicates *one's* possession of *survival*. Choice *A* doesn't do this, so it is incorrect. This is the same for Choice *C*, but that option also takes out the crucial comma after *Jiu-Jitsu*. Choice *D* is incorrect because it changes *play* to *plays*. This disagrees with the plural *Martial arts*, exemplified by having an example of its many forms, *Jiu-Jitsu*. Therefore, *play* is required.

72. C: Choice *C* is the best answer because it most clearly defines the point that the author is trying to make. The original sentence would need a comma after *however* in order to continue the sentence fluidly—but this option isn't available. Choice *B* is close, but this option changes the meaning of the sentence. Therefore, the best alternative is to begin the sentence with *However* and have a comma follow right after it in order to introduce a new idea. The original context is still maintained, but the flow of the language is more streamlined. Thus, Choice *A* is incorrect. Choice *D* would need a comma before and after *however*, so it is also incorrect.

73. A: Choice *A* is the best answer for several reasons. To begin, the section is grammatically correct in using a semicolon to connect the two independent clauses. This allows the two ideas to be connected without separating them. In this context, the semicolon makes more sense for the overall sentence structure and passage as a whole. Choice *B* is incorrect because it forms a run-on. Choice *C* applies a comma in incorrect positions. Choice *D* separates the sentence in a place that does not make sense for the context.

74. D: Choice *D* is the correct answer because it fixes two key issues. First, *their* is incorrectly used. *Their* is a possessive indefinite pronoun and also an antecedent—neither of these fit the context of the sentence, so Choices *A* and *B* are incorrect. What should be used instead is *they're*, which is the contraction of *they are*, emphasizing action or the result of action in this case. Choice *D* also corrects another contraction-related issue with *its*. Again, *its* indicates possession, while *it's* is the contraction of *it is*. The latter is what's needed for the sentence to make sense and be grammatically correct. Thus, Choice *C* is also incorrect.

75. A: Choice *A* is correct because the section contains no errors and clearly communicates the writer's point. Choice *B* is incorrect because it lacks a comma after *philosophy*, needed to link the first clause with the second. Choice *C* also has this issue but additionally alters *its* to *it's*; since *it is* does not make sense in this sentence, this is incorrect. Choice *D* is incorrect because *its* is already plural possessive and does not need an apostrophe on the end.

Math

1. C: The decimal points are lined up, with zeroes put in as needed. Then, the numbers are added just like integers:

$$
\begin{array}{r}
3.40 \\
2.35 \\
+4.00 \\
\hline
9.75
\end{array}
$$

2. A: This problem can be multiplied as 588×32, except at the end, the decimal point needs to be moved three places to the left. Performing the multiplication will give 18,816, and moving the decimal place over three places results in 18.816.

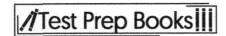

3. E: The fraction is converted so that the denominator is 100 by multiplying the numerator and denominator by 4, to get:

$$\frac{3}{25} = \frac{12}{100}$$

Dividing a number by 100 just moves the decimal point two places to the left, with a result of 0.12.

4. A: Figure out which is largest by looking at the first non-zero digits. Choice *B*'s first non-zero digit is in the hundredths place. The other four all have non-zero digits in the tenths place, so it must be *A*, *C*, *D*, or *E*. Of these, *A* has the largest first non-zero digit.

5. C: 40*N* would be 4000% of *N*. It's possible to check that each of the others is actually 40% of *N*.

6. B: Instead of multiplying these out, the product can be estimated by using $18 \times 10 = 180$. The error here should be lower than 15, since it is rounded to the nearest integer, and the numbers add to something less than 30.

7. C: 85% of a number means multiplying that number by 0.85. So:

$$0.85 \times 20 = \frac{85}{100} \times \frac{20}{1}$$

which can be simplified to:

$$\frac{17}{20} \times \frac{20}{1} = 17$$

8. A: To find the fraction of the bill that the first three people pay, the fractions need to be added, which means finding the common denominator. The common denominator will be 60.

$$\frac{1}{5} + \frac{1}{4} + \frac{1}{3} = \frac{12}{60} + \frac{15}{60} + \frac{20}{60} = \frac{47}{60}$$

The remainder of the bill is:

$$1 - \frac{47}{60} = \frac{60}{60} - \frac{47}{60} = \frac{13}{60}$$

9. B: 30% is 3/10. The number itself must be 10/3 of 6, or:

$$\frac{10}{3} \times 6 = 10 \times 2 = 20$$

10. E: These numbers to improper fractions: $\frac{11}{3} - \frac{9}{5}$. Take 15 as a common denominator:

$$\frac{11}{3} - \frac{9}{5} =: \frac{55}{15} - \frac{27}{15} = \frac{28}{15} = 1\frac{13}{15}$$

.

11. A: Dividing by 98 can be approximated by dividing by 100, which would mean shifting the decimal point of the numerator to the left by 2. The result is 4.2 and rounds to 4.

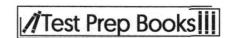

12. B: $4\frac{1}{3} + 3\frac{3}{4} = 4 + 3 + \frac{1}{3} + \frac{3}{4} = 7 + \frac{1}{3} + \frac{3}{4}$

Adding the fractions gives:

$$\frac{1}{3} + \frac{3}{4} = \frac{4}{12} + \frac{9}{12} = \frac{13}{12} = 1 + \frac{1}{12}$$

Thus:

$$7 + \frac{1}{3} + \frac{3}{4} = 7 + 1 + \frac{1}{12} = 8\frac{1}{12}$$

13. C: The average is calculated by adding all six numbers, then dividing by 6. The first five numbers have a sum of 25. If the total divided by 6 is equal to 6, then the total itself must be 36. The sixth number must be 36 − 25 = 11.

14. B: Multiplying by 10^{-3} means moving the decimal point three places to the left, putting in zeroes as necessary.

15. E: $\frac{5}{2} \div \frac{1}{3} = \frac{5}{2} \times \frac{3}{1} = \frac{15}{2} = 7.5$.

16. A: The total fraction taken up by green and red shirts will be:

$$\frac{1}{3} + \frac{2}{5} = \frac{5}{15} + \frac{6}{15} = \frac{11}{15}$$

The remaining fraction is:

$$1 - \frac{11}{15} = \frac{15}{15} - \frac{11}{15} = \frac{4}{15}$$

17. C: If she has used 1/3 of the paint, she has 2/3 remaining. $2\frac{1}{2}$ gallons are the same as $\frac{5}{2}$ gallons.

The calculation is:

$$\frac{2}{3} \times \frac{5}{2} = \frac{5}{3} = 1\frac{2}{3} \text{ gallons}$$

18. B: First, subtract 4 from each side. This yields $6t = 12$. Now, divide both sides by 6 to obtain $t = 2$.

19. A: Each bag contributes $4x + 1$ treats. The total treats will be in the form $4nx + n$ where n is the total number of bags. The total is in the form $60x + 15$, from which it is known $n = 15$.

20. B: If 60% of 50 workers are women, then there are 30 women working in the office. If half of them are wearing skirts, then that means 15 women wear skirts. Since none of the men wear skirts, this means there are 15 people wearing skirts.

21. C: Begin by subtracting 6 from both sides to get $2x = 14$. Dividing both sides by 2 results in $x = 7$.

22. B: To be directly proportional means that $y = mx$. If x is changed from 5 to 20, the value of x is multiplied by 4. Applying the same rule to the y-value, also multiply the value of y by 4. Therefore, $y = 12$.

23. E: Let a be the number of apples and o the number of oranges. Then, the total cost is:

$$2a + 3o = 22$$

while it also known that $a + o = 10$. Using the knowledge of systems of equations, cancel the o variables by multiplying the second equation by -3.

This makes the equation:

$$-3a - 3o = -30$$

Adding this to the first equation, the b values cancel to get $-a = -8$, which simplifies to $a = 8$.

24. A: Finding the roots means finding the values of x when y is zero. The quadratic formula could be used, but in this case it is possible to factor by hand, since the numbers -1 and 2 add to 1 and multiply to -2. So, factor:

$$x^2 + x - 2 = (x - 1)(x + 2) = 0$$

then set each factor equal to zero. Solving for each value gives the values $x = 1$ and $x = -2$.

25. C: To find the y-intercept, substitute zero for x, which gives us:

$$y = 0^{5/3} + (0 - 3)(0 + 1) = 0 + (-3)(1) = -3$$

26. C: Multiply both sides by x to get $x + 2 = 2x$, which simplifies to $-x = -2$, or $x = 2$.

27. D: Denote the width as w and the length as l. Then, $l = 3w + 5$. The perimeter is $2w + 2l = 90$. Substituting the first expression for l into the second equation yields:

$$2(3w + 5) + 2w = 90$$

$$6w + 10 + 2w = 90$$

$$8w = 80$$

$$w = 10$$

Putting this into the first equation, it yields:

$$l = 3(10) + 5 = 35$$

28. A: Lining up the given scores provides the following list: 60, 75, 80, 85, and one unknown. Because the median needs to be 80, it means 80 must be the middle data point out of these five. Therefore, the unknown data point must be the fourth or fifth data point, meaning it must be greater than or equal to 80. The only answer that fails to meet this condition is 60.

29. A: 3.6

Divide 3 by 5 to get 0.6 and add that to the whole number 3, to get 3.6. An alternative is to incorporate the whole number 3 earlier on by creating an improper fraction: $\frac{18}{5}$. Then dividing 18 by 5 to get 3.6.

30. B: 300 miles in 4 hours is $\frac{300}{4}$ = 75 miles per hour. In 1.5 hours, the car will go 1.5×75 miles, or 112.5 miles.

31. A: Mean. An outlier is a data value that is either far above or far below the majority of values in a sample set. The mean is the average of all the values in the set. In a small sample set, a very high or very low number could drastically change the average of the data points. Outliers will have no more of an effect on the median (the middle value when arranged from lowest to highest) than any other value above or below the median. If the same outlier does not repeat, outliers will have no effect on the mode (value that repeats most often).

32. E: Line graph. The scenario involves data consisting of two variables, month and stock value. Box plots display data consisting of values for one variable. Therefore, a box plot is not an appropriate choice. Both line plots and circle graphs are used to display frequencies within categorical data. Neither can be used for the given scenario. Scatter plots compare the values of two variables to see if there are any patterns present. Line graphs display two numerical variables on a coordinate grid and show trends among the variables.

33. D: $\frac{1}{12}$. The probability of picking the winner of the race is $\frac{1}{4}$, or:

$$\left(\frac{number\ of\ favorable\ outcomes}{number\ of\ total\ outcomes}\right)$$

Assuming the winner was picked on the first selection, three horses remain from which to choose the runner-up (these are dependent events). Therefore, the probability of picking the runner-up is $\frac{1}{3}$. To determine the probability of multiple events, the probability of each event is multiplied:

$$\frac{1}{4} \times \frac{1}{3} = \frac{1}{12}$$

34. B: This inequality can be seen with the use of a number line. $\frac{3}{7}$ is close to $\frac{1}{2}$.

$\frac{5}{6}$ is close to 1, but less than 1, and $\frac{8}{7}$ is greater than 1. Therefore, $\frac{3}{7}$ is less than $\frac{5}{6}$.

35. D: 104,165

Set up the problem and add each column, starting on the far right (ones). Add, carrying anything over 9 into the next column to the left. Solve from right to left.

36. E: 6.630

Set up the problem, with the larger number on top and numbers lined up at the decimal. Add, carrying anything over 9 into the next column to the left. Solve from right to left.

37. B: 148.97

Set up the problem, with the larger number on top and numbers lined up at the decimal. Insert 0 in any blank spots to the right of the decimal as placeholders. Add, carrying anything over 9 into the next column to the left.

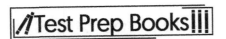

38. C: Each number in the sequence is adding one more than the difference between the previous two.

For example, $10 - 6 = 4, 4 + 1 = 5$.

Therefore, the next number after 10 is $10 + 5 = 15$.

Going forward, $21 - 15 = 6, 6 + 1 = 7$. The next number is $21 + 7 = 28$.

Therefore, the difference between numbers is the set of whole numbers starting at 2: 2, 3, 4, 5, 6, 7....

39. C: $\frac{4}{3}$

Set up the problem and find a common denominator for both fractions:

$$\frac{14}{33} + \frac{10}{11}$$

Multiply each fraction across by 1 to convert to a common denominator:

$$\frac{14}{33} \times \frac{1}{1} + \frac{10}{11} \times \frac{3}{3}$$

Once over the same denominator, add across the top. The total is over the common denominator:

$$\frac{14 + 30}{33} = \frac{44}{33}$$

Reduce by dividing both numerator and denominator by 11:

$$\frac{44 \div 11}{33 \div 11} = \frac{4}{3}$$

40. B: 128

This question involves the percent formula:

$$\frac{32}{x} = \frac{25}{100}$$

We multiply the diagonal numbers, 32 and 100, to get 3,200. Dividing by the remaining number, 25, gives us 128.

The percent formula does not have to be used for a question like this. Since 25% is ¼ of 100, you know that 32 needs to be multiplied by 4, which yields 128.

41. D: 110,833

Set up the problem, with the larger number on top. Begin subtracting with the far right column (ones). Borrow 10 from the column to the left, when necessary.

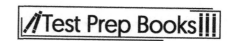

42. B: This is a division problem because the original amount needs to be split up into equal amounts. The mixed number $11\frac{1}{2}$ should be converted to an improper fraction first:

$$11\frac{1}{2} = \frac{(11 \times 2) + 1}{2} = \frac{23}{2}$$

Carey needs to determine how many times $\frac{23}{2}$ goes into 184. This is a division problem:

$$184 \div \frac{23}{2} = ?$$

The fraction can be flipped, and the problem turns into the multiplication:

$$184 \times \frac{2}{23} = \frac{368}{23}$$

This improper fraction can be simplified into 16 because $368 \div 23 = 16$. The answer is 16 lawn segments.

43. B: 648.77

Set up the problem, with the larger number on top and numbers lined up at the decimal. Insert 0 in any blank spots to the right of the decimal as placeholders. Begin subtracting with the far right column. Borrow 10 from the column to the left, when necessary.

44. A: The additive and subtractive identity is 0. When added or subtracted to any number, 0 does not change the original number.

45. E: 100 cm is equal to 1 m. 1.3 divided by 100 is 0.013. Therefore, 1.3 cm is equal to 0.013 m. Because 1 cm is equal to 10 mm, 1.3 cm is equal to 13 mm.

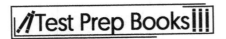

46. D: 725

Set up the division problem.

1. 4 | 1   0   1   5

Move the decimal over one place to the right in both numbers.

1   4 | 1   0   1   5   0

14 does not go into 1 or 10 but does go into 101 so start there.

```
                        7   2   5
    1   4 | 1   0   1   5   0
          -   9   8
                  3   5
              -   2   8
                      7   0
                  -   7   0
                          0
```

The result is 725.

47. C: One apple/orange pair costs $3 total. Therefore, Jan bought $\frac{90}{3} = 30$ total pairs, and hence, she bought 30 oranges.

48. B: 2.48

Set up the problem, with the larger number on top. Multiply as if there are no decimal places. Add the answer rows together. Count the number of decimal places that were in the original numbers ($1 + 1 = 2$).

Place the decimal 2 places to the right for the final solution.

49. B: 99.35

Set up the problem, with the larger number on top. Multiply as if there are no decimal places. Add the answer rows together. Count the number of decimal places that were in the original numbers (2).

Place the decimal in that many spots from the right for the final solution.

50. B: Each hour on the clock represents 30 degrees. For example, 3:00 represents a right angle. Therefore, 5:00 represents 150 degrees.

51. D: $27\frac{7}{22}$

Set up the division problem.

4   4 | 1   2   0   2

44 does not go into 1 or 12 but will go into 120 so start there.

```
                    2    7
   4   4 | 1   2    0    2
       -     8    8
             3    2    2
       -     3    0    8
                  1    4
```

The answer is $27\frac{14}{44}$.

Reduce the fraction for the final answer.

$$27\frac{7}{22}$$

52. E: The formula for the volume of a box with rectangular sides is the length times width times height, so $5 \times 6 \times 3 = 90$ cubic feet.

53. D: 270

Set up the division problem.

```
   2.   6 | 7    0    2
```

Move the decimal over one place to the right in both numbers.

```
   2    6 | 7    0    2    0
```

26 does not go into 7 but does go into 70 so start there.

```
                    2    7    0
   2    6 | 7    0    2    0
       -  5    2
          1    8    2
       -  1    8    2
                    0
```

The result is 270.

54. D: First, the train's journey in the real world is $3 \times 50 = 150$ miles. On the map, 1 inch corresponds to 10 miles, so there is $\frac{150}{10} = 15$ inches on the map.

55. B: The total trip time is $1 + 3.5 + 0.5 = 5$ hours. The total time driving is $1 + 0.5 = 1.5$ hours. So, the fraction of time spent driving is 1.5/5 or 3/10. To get the percentage, convert this to a fraction out of 100. The numerator and denominator are multiplied by 10, with a result of 30/100. The percentage is the numerator in a fraction out of 100, so 30%.

56. B: The formula for the volume of a cylinder is $\pi r^2 h$, where r is the radius and h is the height. The diameter is twice the radius, so these barrels have a radius of 1 foot. That means each barrel has a

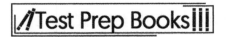

volume of $\pi \times 1^2 \times 3 = 3\pi$ cubic feet. Since there are three of them, the total is $3 \times 3\pi = 9\pi$ cubic feet.

57. A: The tip is not taxed, so he pays 5% tax only on the $10. 5% of $10 is $0.05 \times 10 = \$0.50$. Add up $\$10 + \$2 + \$0.50$ to get $12.50.

58. B: 15,412

Set up the problem and add each column, starting on the far right (ones). Add, carrying anything over 9 into the next column to the left. Solve from right to left.

59. A: Using the order of operations, multiplication and division are computed first from left to right. Multiplication is on the left; therefore, multiplication should be performed first.

60. D: A factor of 36 is any number that can be divided into 36 and have no remainder. $36 = 36 \times 1, 18 \times 2, 9 \times 4,$ and 6×6. Therefore, it has 7 unique factors: 36, 18, 9, 6, 4, 2, and 1.

Reading

1. D: It emphasizes Mr. Utterson's anguish in failing to identify Hyde's whereabouts. Context clues indicate that Choice *D* is correct because the passage provides great detail of Mr. Utterson's feelings about locating Hyde. Choice *A* does not fit because there is no mention of Mr. Lanyon's mental state. Choice *B* is incorrect; although the text does make mention of bells, Choice *B* is not the *best* answer overall. Choice *C* is incorrect because the passage clearly states that Mr. Utterson was determined, not unsure.

2. A: In the city. The word *city* appears in the passage several times, thus establishing the location for the reader.

3. B: It scares children. The passage states that the Juggernaut causes the children to scream. Choices *A* and *D* don't apply because the text doesn't mention either of these instances specifically. Choice *C* is incorrect because there is nothing in the text that mentions space travel.

4. B: To constantly visit. The mention of *morning*, *noon*, and *night* make it clear that the word *haunt* refers to frequent appearances at various times. Choice *A* doesn't work because the text makes no mention of levitating. Choices *C* and *D* are not correct because the text makes mention of Mr. Utterson's anguish and disheartenment because of his failure to find Hyde but does not make mention of Mr. Utterson's feelings negatively affecting anyone else.

5. D: This is an example of alliteration. Choice *D* is the correct answer because of the repetition of the *L*-words. Hyperbole is an exaggeration, so Choice *A* doesn't work. No comparison is being made, so no simile or juxtaposition is being used, thus eliminating Choices B and C.

6. D: The speaker intends to continue to look for Hyde. Choices *A* and *B* are not possible answers because the text doesn't refer to any name changes or an identity crisis, despite Mr. Utterson's extreme obsession with finding Hyde. The text also makes no mention of a mistaken identity when referring to Hyde, so Choice *C* is also incorrect.

7. A: The author's attitude toward the main subject can be described as *intrigue*. Although this is fiction and we are seeing the passage through the eyes of a character, the author still is in control of word choice and tone. *Intrigue* means to arouse curiosity, so we are confronted with words and phrases such

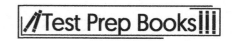

as "besieged by questions," "digging," "imagination," and "haunt." Choice *B*, elation, means joy. Choice *C*, animosity, means strong dislike. Choice *D*, rigidity, means stiff or unyielding.

8. B: Mr. Utterson is struggling with the mystery surrounding a dark figure and the terrible crimes he commits. As Mr. Utterson tosses and turns in bed in the long paragraph, we see him wanting to discern the figure's face as he imagines him committing the crimes, but Mr. Utterson has no idea who the figure is.

9. C: Choice *C* is the best answer because of the chronological aspect of the passage. By the transition "From that time forward," we know that Mr. Utterson haunted storefronts *after* his night visions of the mysterious figure, and we also see that Mr. Utterson's dialogue at the very end is a promise to find "Mr. Hyde," whoever he may be.

10. C: The passage is in third person limited, which means we see the thoughts of one character only by the use of the pronouns "he" or "she." First person is characterized by the use of "I." Second person is characterized by the use of "you." Third person omniscient is when we see the thoughts of all the characters in the story, and the author uses the pronouns "he" or "she."

11. A: The purpose is to inform the reader about what assault is and how it is committed. Choice *B* is incorrect because the passage does not state that assault is a lesser form of lethal force, only that an assault can use lethal force, or alternatively, lethal force can be utilized to counter a dangerous assault. Choice *C* is incorrect because the passage is informative and does not have a set agenda. Finally, Choice *D* is incorrect because although the author uses an example in order to explain assault, it is not indicated that this is the author's personal account.

12. C: If the man being attacked in an alley by another man with a knife used self-defense by lethal force, it would not be considered illegal. The presence of a deadly weapon indicates mal-intent and because the individual is isolated in an alley, lethal force in self-defense may be the only way to preserve his life. Choices *A* and *B* can be ruled out because in these situations, no one is in danger of immediate death or bodily harm by someone else. Choice *D* is an assault and does exhibit intent to harm, but this situation isn't severe enough to merit lethal force; there is no intent to kill.

13. B: As discussed in the second passage, there are several forms of assault, like assault with a deadly weapon, verbal assault, or threatening posture or language. Choice *A* is incorrect because the author does mention what the charges are on assaults; therefore, we cannot assume that they are more or less than unnecessary use of force charges. Choice *C* is incorrect because anyone is capable of assault; the author does not state that one group of people cannot commit assault. Choice *D* is incorrect because assault is never justified. Self-defense resulting in lethal force can be justified.

14. D: The use of lethal force is not evaluated on the intent of the user, but rather on the severity of the primary attack that warranted self-defense. This statement most undermines the last part of the passage because it directly contradicts how the law evaluates the use of lethal force. Choices *A* and *B* are stated in the paragraph, so they do not undermine the explanation from the author. Choice *C* does not necessarily undermine the passage, but it does not support the passage either. It is more of an opinion that does not offer strength or weakness to the explanation.

15. C: An assault with deadly intent can lead to an individual using lethal force to preserve their well-being. Choice *C* is correct because it clearly establishes what both assault and lethal force are and gives the specific way in which the two concepts meet. Choice *A* is incorrect because lethal force doesn't necessarily result in assault. This is also why Choice *B* is incorrect. Not all assaults would necessarily be

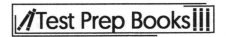
life-threatening to the point where lethal force is needed for self-defense. Choice *D* is compelling but ultimately too vague; the statement touches on aspects of the two ideas but fails to present the concrete way in which the two are connected to each other.

16. A: Both passages open by defining a legal concept and then continue to describe situations in order to further explain the concept. Choice *D* is incorrect because while the passages utilize examples to help explain the concepts discussed, the author doesn't indicate that they are specific court cases. It's also clear that the passages don't open with examples, but instead, they begin by defining the terms addressed in each passage. This eliminates Choice *B,* and ultimately reveals Choice *A* to be the correct answer. Choice *A* accurately outlines the way both passages are structured. Because the passages follow a nearly identical structure, the Choice *C* can easily be ruled out.

17. C: Intent is very important for determining both lethal force and assault; intent is examined in both parties and helps determine the severity of the issue. Choices *A* and *B* are incorrect because it is clear in both passages that intent is a prevailing theme in both lethal force and assault. Choice *D* is compelling, but if a person uses lethal force to defend himself or herself, the intent of the defender is also examined in order to help determine if there was excessive force used. Choice *C* is correct because it states that intent is important for determining both lethal force and assault, and that intent is used to gauge the severity of the issues. Remember, just as lethal force can escalate to excessive use of force, there are different kinds of assault. Intent dictates several different forms of assault.

18. B: The example is used to demonstrate a single example of two different types of assault, then adding in a third type of assault to the example's conclusion. The example mainly provides an instance of "threatening body language" and "provocative language" with the homeowner gesturing threats to his neighbor. It ends the example by adding a third type of assault: physical strikes. This example is used to show the variant nature of assaults. Choice *A* is incorrect because it doesn't mention the "physical strike" assault at the end and is not specific enough. Choice *C* is incorrect because the example does not say anything about the definition of lethal force or how it might be altered. Choice *D* is incorrect, as the example mentions nothing about cause and effect.

19. C: The word *apprehension* most nearly means fear. The passage indicates that "assault is the unlawful attempt of one person to apply fear/anxiety on another individual by an imminent threat." The creation of fear in another individual seems to be a property of assault.

20. D: Choice *D* is the best answer, "The concept of lethal force as a last resort and the point at which it can cross a line from defense to manslaughter." The last paragraph of the first passage states what the term "last resort" means and how it's distinguished in the eyes of the law.

21. A: The word *patronage* most nearly means *auspices,* which means *protection* or *support.* Choice *B, aberration,* means *deformity* and does not make sense within the context of the sentence. Choice *C, acerbic,* means *bitter* and also does not make sense in the sentence. Choice *D, adulation,* is a positive word meaning *praise,* and thus does not fit with the word *condescending* in the sentence.

22. D: *Working man* is most closely aligned with Choice *D, bourgeois.* In the context of the speech, the word *bourgeois* means *working* or *middle class.* Choice *A, Plebeian,* does suggest *common people;* however, this is a term that is specific to ancient Rome. Choice *B, viscount,* is a European title used to describe a specific degree of nobility. Choice *C, entrepreneur,* is a person who operates their own business.

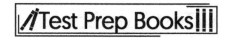

23. C: In the context of the speech, the term *working man* most closely correlates with Choice *C*, "A working man is someone who works for wages among the middle class." Choice *A* is not mentioned in the passage and is off-topic. Choice *B* may be true in some cases, but it does not reflect the sentiment described for the term *working man* in the passage. Choice *D* may also be arguably true. However, it is not given as a definition but as *acts* of the working man, and the topics of *field, factory,* and *screen* are not mentioned in the passage.

24. D: *Enterprise* most closely means *cause*. Choices *A, B,* and *C* are all related to the term *enterprise.* However, Dickens speaks of a *cause* here, not a company, courage, or a game. "He will stand by such an enterprise" is a call to stand by a cause to enable the working man to have a certain autonomy over his own economic standing. The very first paragraph ends with the statement that the working man "shall . . . have a share in the management of an institution which is designed for his benefit."

25. B: The speaker's salutation is one from an entertainer to his audience and uses the friendly language to connect to his audience before a serious speech. Recall in the first paragraph that the speaker is there to "accompany [the audience] . . . through one of my little Christmas books," making him an author there to entertain the crowd with his own writing. The speech preceding the reading is the passage itself, and, as the tone indicates, a serious speech addressing the "working man." Although the passage speaks of employers and employees, the speaker himself is not an employer of the audience, so Choice *A* is incorrect. Choice *C* is also incorrect, as the salutation is not used ironically, but sincerely, as the speech addresses the well-being of the crowd. Choice *D* is incorrect because the speech is not given by a politician, but by a writer.

26. B: Choice *A* is incorrect because that is the speaker's *first* desire, not his second. Choices *C* and *D* are tricky because the language of both of these is mentioned after the word *second.* However, the speaker doesn't get to the second wish until the next sentence. Choices *C* and *D* are merely prepositions preparing for the statement of the main clause, Choice B, for the working man to have a say in his institution, which is designed for his benefit.

27. C: The speaker's tone can best be described as *confident and informed.* The speaker addresses the audience as "My good friends," and says, "I have no fear of being misunderstood," which implies confidence. Additionally, the speaker's knowledge of the proposal and topic can be seen in the text as well, especially in the second paragraph.

28. D: To provide credibility to the working man and share confidence in their ability to take on responsibilities if they are compensated appropriately. The speaker provides credibility by saying "he will stand by such an enterprise with the utmost of his patience," and displays their responsibilities by saying "he will feel his responsibility like an honest man."

29. A: The speaker says to "Erect in Birmingham a great Education Institution, properly educational." Choice *B* is close, but the speaker uses the name "Temple of Concord" in the passage as a metaphor, so this is incorrect. The other two choices aren't mentioned in the passage.

30. B: The word *antagonism* most nearly means opposition. Choice *A*, *conformity*, is the opposite of antagonism. Choice *C*, *affluence*, means abundance. Choice *D*, *scarcity*, means being deficient in something.

31. D: The use of "I" could serve to have a "hedging" effect, allow the reader to connect with the author in a more personal way, and cause the reader to empathize more with the egrets. However, it doesn't distance the reader from the text, making Choice *D* the answer to this question.

32. C: The quote provides an example of a warden protecting one of the colonies. Choice *A* is incorrect because the speaker of the quote is a warden, not a hunter. Choice *B* is incorrect because the quote does not lighten the mood but shows the danger of the situation between the wardens and the hunters. Choice *D* is incorrect because there is no humor found in the quote.

33. D: A *rookery* is a colony of breeding birds. Although *rookery* could mean Choice *A*, houses in a slum area, it does not make sense in this context. Choices *B* and *C* are both incorrect, as this is not a place for hunters to trade tools or for wardens to trade stories.

34. B: An important bird colony. The previous sentence is describing "twenty colonies" of birds, so what follows should be a bird colony. Choice *A* may be true, but we have no evidence of this in the text. Choice *C* does touch on the tension between the hunters and wardens, but there is no official "Bird Island Battle" mentioned in the text. Choice *D* does not exist in the text.

35. D: To demonstrate the success of the protective work of the Audubon Association. The text mentions several different times how and why the association has been successful and gives examples to back this fact. Choice *A* is incorrect because although the article, in some instances, calls certain people to act, it is not the purpose of the entire passage. There is no way to tell if Choices *B* and *C* are correct, as they are not mentioned in the text.

36. C: To have a better opportunity to hunt the birds. Choice *A* might be true in a general sense, but it is not relevant to the context of the text. Choice *B* is incorrect because the hunters are not studying lines of flight to help wardens, but to hunt birds. Choice *D* is incorrect because nothing in the text mentions that hunters are trying to build homes underneath lines of flight of birds for good luck.

37. C: The passage states in the third paragraph that a year before, "a number of egrets were seen in eastern Massachusetts." Florida is mentioned in the passage as a place where bird colonies reside. The other two locations are not mentioned in the passage.

38. A: The word *commissioned* most nearly means *appointed*. Choice *B*, *compelled*, means forced. Choice *C*, *beguiled*, means entertained. Choice *D*, fortified, means defended.

39. D: Several hidden colonies were found which prompted Dutcher and Pearson to protect them. This information is presented in the first paragraph starting with the sentence "Two years ago." The other answer choices are current to the passage.

40. C: *In transit* means on the way to the destination. Choice *A* is the definition of *travelling*. Choice *B* is the definition of *delegate*. Choice *D* is the definition of *provoke*.

Science

1. A: The hypothesis is the sentence that describes what the scientist wants to research with a conclusive expected finding. Choice *A* describes how she believes sunlight will affect plant growth. Choice *B* includes details about the experiment. Choice *C* is not a conclusive theory. Choice *D* describes the data that she found after conducting the experiment.

2. C: Looking at Figure 1, four experimental groups are shown on the graph for which data were collected: plants that received 1 hour of sunlight, 3 hours of sunlight, 5 hours of sunlight, and 7 hours of sunlight. Choices *A* and *B* could be describing two of the experimental groups and how much sunlight they received. Choice *D* describes how many days' data was collected.

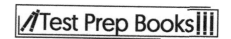

3. B: After the data was collected, it was compiled into a line graph. The data points were collected, and then a line was drawn between the points. Data is represented by horizontal or vertical bars in bar graphs, Choice *A*. Pie charts are circular charts, with the data being represented by different wedges of the circle, Choice *C*. Pictograms use pictures to describe their subject, Choice *D*.

4. A: Looking at Figure 2, the sun provides light energy that drives forward the process of photosynthesis, which is how plants make their own source of energy and nutrients. Choices *B* and *C* are found in the environment around the plants. They combine with light energy to make the photosynthesis reaction work. Choice *D* is a product of photosynthesis.

5. D: Looking at the Figure 1, the experimental group that received 7 hours of sunlight every day grew taller than any of the other groups that received less sunlight per day. Therefore, it is reasonable to conclude that more sunlight makes plants grow bigger. Choice *A* is not a reasonable conclusion because it did not have the tallest plants. The scientist decided to measure the plants only for 11 days, but that does not describe a conclusion for the experiment, Choice *B*. Choice *C* is the opposite of the correct conclusion and does not have evidence to support it.

6. B: The atomic mass of a molecule can be found by adding the atomic mass of each component together. Looking at Figure 2, the atomic mass of each element is found below its symbol. The atomic mass of Na is 23, Choice *A*, and the atomic mass of Cl is 35.5, Choice *C*. The sum of those two components is 58.5, Choice *B*. Choice *D* is equal to two Cl atoms joined together.

7. D: Figure 1 shows the trends of the periodic table. Looking at the black arrows representing electronegativity, it is shown that electronegativity increases going towards the top row of the table and also increases going towards the right columns of the table. Therefore, the most electronegative element would be found in the top right corner of the table, which is where the element Helium is found. Choices *A* and *B* are found at the bottom of the table. Choice *D* is found on the left side of the table.

8. A: Looking at Figure 1, the ionization energy increases as it goes right and up on the periodic table. Of the choices given, potassium (K) is found in the first column, fourth row. Each of the other choices is found above or to the right of potassium, so they would tend to have higher ionization energy. Since we're looking for the element with the lowest ionization energy, Choice *A*, potassium, is correct.

9. D: The atomic number of an element represents the number of protons. Looking at Figure 2, the atomic number is located at the top of the box, above the element's symbol. Hydrogen (H) has an atomic number of 1 and has the least number of protons of any other element in the periodic table. Radon (Rn), Choice *A*, has 86 protons. Boron (B), Choice *B*, has 5 protons. Nitrogen (N), Choice *C*, has 7 protons.

10. B: Looking at Figure 3, the elements are color coded in periods and groups according to their similar properties. Noble gases are located in the right most column of the table. Radon (Rn) is the only one of the element choices marked as a noble gas and would be the right choice for Scientist A. Nitrogen (N) and Boron (B), Choices *A* and *D*, are nonmetals. Copper (Cu), Choice *C*, is a transition metal.

11. B: Looking at Figures 1 and 2, crossbreeding experiment #3 in round #1 produces plants that are completely recessive and would have white flowers. Choices *A* and *C*, crossbreeding experiments #1 and 2, respectively, only produce flowers with a dominant allele present, making red flowers. Choice *D* does not have any recessive alleles, so white flowers are not a possibility.

12. C: Looking at Figure 2, which represents the number of plants that were produced from each crossbreeding experiment, it can be seen that only 2 plants produced white flowers out of 12 plants total, 4 from each experiment. To find the percentage, divide 2 by 12 and multiply by 100. The result is 16.7%. Choice *A* is the total number of plants that were produced. Choice *B* represents the percentage of white flowers in experiment #3 alone.

13. B: In a Punnett Square, each box represents one allele from each of the parent's genes. To find the genetic makeup of the second parent, take out the allele that was contributed from the first parent. Here, the first parent contributed a recessive allele, a, to each offspring. In the top row, that leaves a dominant allele, A, and in the bottom row, that leaves a recessive allele, a. Therefore, the genetic makeup of the second parent is Aa.

14. D: Crossbreeding the plants with only recessive alleles will result in 100% white flowering plants. All four offspring have white flowering plants. Choice *B* gives 100% red flowering plants. Choice *C* gives 25%, 1 out of 4 plants, with white flowers.

15. A: The observation step of the scientific method involves using your senses to identify the results of the experiment. In this case, the experiment depended on identifying the color of the flowers. This was done using sight. If the experiment had involved different scents produced by the flowers, Choice *B* would have worked. If it has involved different textures of the flowers, Choice *C* would have worked. Flowers generally do not make any noise, so Choice *D* would not have been useful.

16. D: The process of photosynthesis requires carbon dioxide and water to combine with sunlight to produce glucose, which is used as an energy source by plants. The forest floor does not get a lot of sunlight since it is shaded by the growth of so many trees and plants in the rainforest. Carbon, Choice *A*, is available through the air. Plants expel carbon dioxide. Water, Choice *B*, is abundant in the humid climate of the rainforest. Oxygen, Choice *C*, is always available in the Earth's atmosphere.

17. B: Central America is one of the five major areas of the world that has a rainforest. Looking at Figure 2, it can be seen that the southern countries of Central America contain rainforests. Comparing this map to the map of Central America, it is clear that Panama is a country that has rainforests. Choices *A*, *C*, and *D* are not found in Central America.

18. C: Birds fly above the trees of the rainforest the most. There, they have unobstructed skies, unlike the dense growth of the trees and plants in the other layers of the rainforest.

19. D: The understory layer is the third layer from the top of the rainforest. It does not receive much sunlight, so the plants need to grow large leaves to absorb as much sunlight as possible. Giant taro leaves would grow well in this layer since they have large leaves. The emergent layer, Choice *A*, gets plenty of sunlight since it is the topmost layer. The canopy layer, Choice *B*, receives enough sunlight for plants to grow without needing to increase their leaf size. The forest floor, Choice *C*, does not receive sunlight, and plants generally do not grow here.

20. A: Rainforests have warm and wet climates. They do not have long dry seasons and tend to have temperate temperatures. The giant water lily is ideal for the rainforest because it can grow large leaves and needs a wet environment to grow in. They grow in the shallow basins of rainforest rivers. Choices *B* and *C* need dry environments. Choice *D* needs a very hot environment, which is not characteristic of rainforests.

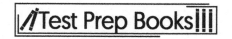

21. B: Solar eclipses should not be looked at directly. The rays of the Sun do not seem as bright as normal but can still cause damage to the eyes. A pinhole camera facing away from the eclipse allows the viewer to see a reflection of the eclipse instead of the actual eclipse. Choices *A, C,* and *D* all require looking directly at the solar eclipse.

22. A: When the moon comes between the Earth and Sun, a solar eclipse occurs. If the sun is far enough away and is completely blocked by the moon, it is a total solar eclipse. If it is only partially blocked by the moon, it is a partial solar eclipse, Choice *D*. A lunar eclipse occurs when the moon is on the opposite side of the Earth as the Sun and the Sun creates a shadow of the Earth on the moon, so that the moon becomes completely dark, Choice *C*, or partially dark, Choice *B*.

23. C: During a lunar eclipse, the Sun and moon are on opposite sides of the Earth. They line up so that the Sun's light that normally illuminates the moon is blocked by the Earth. This causes the moon to become dim. Sunlight can still be seen, Choice *A*, and the Earth does not become dark, Choices *B* and *D*.

24. B: The moon does not produce harmful light rays that can damage the eyes, so lunar eclipses can be viewed directly. A telescope would allow the lunar eclipse to be magnified and seen more clearly. During a solar eclipse, the Sun's rays appear to be dim and easy to see directly but they are still harmful to the eyes.

25. B: Solar eclipses are viewed during the daytime because they involve viewing the Sun while it is out during normal daytime hours. Lunar eclipses, Choices *C* and *D*, are viewed at nighttime when the moon is in the sky during its normal hours. The moon is normally illuminated by the Sun that is on the other side of the Earth. When the Sun is on the other side of the earth, it is nighttime for people looking at the moon.

26. B: A cladogram is a diagram that organizes proposed ancestral relations based on the development of physical features. A branching point would be seen on the cladogram where the development of lungs was noted. A phylogenetic tree, Choice *A*, does not note phenotypic features on it. Punnett squares, Choice *C*, are used to determine the possible genetic makeup of offspring and are not related to evolution. Photographs, Choice *D*, may reveal species that look alike but would not reveal if they truly had a common ancestor.

27. A: According to Figure 3, human arms and whale fins are homologous structures that were derived from a common ancestor. They have anatomical similarities, although their function is not the same. They have different numbers of bones, so Choice *B* is incorrect. Since they are proposed to be developed from a common ancestor, they are not analogous features, Choice *C*. Whales have blubber covering their bodies and not layered skin like humans, so Choice *D* is incorrect.

28. D: According to the phylogenetic tree in Figure 1, the common ancestor of the striped skunk and European otter is the one that is noted before they branch into separate lineages, which is Mustelidae. Mephitis, Choice *A*, is the genus for only the striped skunk. Felidae, Choice *B*, and Canidae, Choice *C*, are completely different branches of the Carnivora order than the one that leads to the striped skunk and European otter.

29. A: According to Figure 2, the common trait that is listed on the branch of the cladogram that leads to lizards and birds is seeing UV light. They also have the common traits listed on the main branch of the cladogram before their lineages are branched off, which are vertebrae, lungs, and amniotic eggs. Perch and flounder branch from the main common ancestor and develop spiny-rayed fins, Choice *B*. Lobsters and spiders branch from the main common ancestor and develop the ability to molt an exoskeleton, Choice *C*.

30. C: Domestic cats and wolves are proposed to be related at the point where they share a common line before any branching occurs to separate their lineages. Figure 1 shows this as Carnivora, which is noted as the Order on the left side of the figure.

31. D: The color reagent is attached to the secondary antibody. It is released only when the secondary antibody attaches to the activated primary antibody. The antigen, primary antibody, and capture antibody, Choices *A*, *B*, and *C*, do not have any color reagent attached to them, so only the secondary antibody can cause the color reaction.

32. B: The color reagent is attached to the secondary antibody. If more antigen is present, more primary and secondary antibody will be attached to it and more color reagent will be released. Row 1 has the darkest green color of all the samples tested in the plate in Figure 2.

33. C: Looking at the graph in Figure 3, the highest amount of IL-1β is found in the spleen. IL-1β is a marker of inflammation and indicates that the spleen had the most inflammation of the areas tested. Serum, Choice *A*, had no IL-1β in the sample. Plasma, Choice *B*, had the second highest amount of IL-1β in the sample, and bone marrow, Choice *D*, had the second lowest amount of IL-1β.

34. B: Looking at the diagram in Figure 1, the antigen is located between the capture antibody and the primary antibody. The capture antibody keeps the antigen attached to the surface of the plate. The primary antibody recognizes the specific antigen. The secondary antibody generally recognizes the primary antibody is not specific to the antigen.

35. A: ELISAs are used to analyze specific substances within a larger sample. The antibodies used in an ELISA are designed specifically for a particular antigen. Sandwich ELISAs are generally used to quantify one antigen and not all substances in a larger sample, making Choice *B* incorrect. When used to quantify an antigen, the antibodies need to already be developed and able to detect the antigen, making Choice *D* incorrect.

36. A: Stabilizing selection occurs when an intermediate phenotype is favored over two extreme phenotypes. In Scenario 1, the mice develop an intermediate colored fur so that they can blend in with the rocks in their environment. Developing one or both extremes, Choices *B* and *C*, would make them more visible to predators. Color selection, Choice *D*, is not a type of natural selection.

37. D: Directional selection occurs when one extreme of a phenotype is favored. In Scenario 2, large beaks are favored over medium- or small-sized beaks. The large beaks help the finches break up the tough seeds that became abundant after the drought. Finches with medium and small beaks had trouble breaking up the large seeds and did not survive as well as those with large beaks.

38. B: Small-beaked finches had trouble breaking up the large seeds after the drought, and therefore could not gain enough nutrition for survival. Natural selection is based on the idea that the individuals who adapt to their environment in the best way are the ones that have enhanced survival and reproduction. Finches with large beaks were most able to adapt to the large seeds and continue with their regular feeding schedule. There was no evidence of Choices *A* or *C* in the passage. Two extremes can be selected by natural selection in disruptive selection but that was not the case here, Choice *D*.

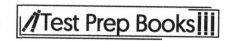
39. C: Disruptive selection occurs when both extremes of a phenotype are selected. In Cameroon, the finches had both large and small beaks, but did not survive well with medium beaks. If medium beaks were selected, it would have been stabilizing selection, Choice *A*. If only one of the extremes had been favored, it would have been directional selection, Choice *B*.

40. D: Natural selection is the idea that individuals are selected to survive and reproduce based on their ability to adapt to the environment. Their phenotypes are advantageous for survival and reproduction over those of other individuals. It is solely based on the phenotype of the individual, not the genotype, Choice *A*. Extreme phenotypes, Choice *B*, may be selected but are not always the most advantageous. It occurs all time, not just in extreme weather conditions, such as a drought, Choice *C*.

ACT Practice Test #2

English

Questions 1–9 are based on the following passage:

The name "Thor" has always been associated with great power. (1) <u>Arguably, Norse Mythologies most</u> popular and powerful god is Thor of the Aesir. My first experience of Thor was not like most of today's generation. I grew up reading Norse mythology where (2) <u>Thor wasn't a comic book superhero, but even mightier.</u> There are stories of Thor destroying mountains, (3) <u>defeating scores of giants and lifting up the world's largest creature the Midgard Serpent.</u> But always, Thor was a protector.

Like in modern comics and movies, Thor was the god of thunder and wielded (4) <u>the hammer Mjolnir however there are several differences</u> between the ancient legend and modern hero. (5) <u>For example, Loki, the god of mischief, isn't Thor's brother.</u> Loki is actually Thor's servant, but this doesn't stop the trickster from causing chaos, chaos that Thor has to then quell. In all of his incarnations, Thor is a god that reestablishes order by tempering the chaos around him. (6) <u>This is also symbolized in his prized weapon Mjolnir a magic hammer.</u> A hammer is both a weapon and a (7) <u>tool, but why would a god favor a seemingly everyday object?</u>

A hammer is used to shape metal and create change. The hammer tempers raw iron, (8) <u>ore that is in an chaotic state of impurities and shapelessness,</u> to create an item of worth. Thus, a hammer is in many ways a tool that brings a kind of order to the world—like Thor. Hammers were also tools of everyday people, which further endeared Thor to the common man. Therefore, it's no surprise that Thor remains an iconic hero to this day.

I began thinking to myself, why is Thor so prominent in our culture today even though many people don't follow the old religion? (9) <u>Well the truth is that every culture throughout time, including ours,</u> needs heroes. People need figures in their lives that give them hope and make them aspire to be great. We need the peace of mind that chaos will eventually be brought to order and that good can conquer evil. Thor was a figure of hope and remains so to this day.

1. Which of the following would be the best choice for this sentence (reproduced below)?

(1) <u>Arguably, Norse Mythologies most</u> popular and powerful god is Thor of the Aesir.

a. NO CHANGE
b. Arguably Norse Mythologies most
c. Arguably, Norse mythology's most
d. Arguably, Norse Mythology's most

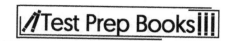

2. Which of the following would be the best choice for this sentence (reproduced below)?

I grew up reading Norse mythology where (2) <u>Thor wasn't a comic book superhero, but even mightier.</u>

   a. NO CHANGE
   b. Thor wasn't a comic book superhero. He was even mightier.
   c. Thor wasn't a comic book superhero but even mightier.
   d. Thor wasn't a comic book superhero, he was even mightier.

3. Which of the following would be the best choice for this sentence (reproduced below)?

There are stories of Thor destroying mountains, (3) <u>defeating scores of giants and lifting up the world's largest creature the Midgard Serpent.</u>

   a. NO CHANGE
   b. defeating scores of giants, and lifting up the world's largest creature, the Midgard Serpent.
   c. defeating scores of giants, and lifting up the world's largest creature the Midgard Serpent.
   d. defeating scores, of giants, and lifting up the world's largest creature the Midgard Serpent.

4. Which of the following would be the best choice for this sentence (reproduced below)?

Like in modern comics and movies, Thor was the god of thunder and wielded (4) <u>the hammer Mjolnir however there are several differences</u> between the ancient legend and modern hero.

   a. NO CHANGE
   b. the hammer Mjolnir, however there are several differences
   c. the hammer Mjolnir. However there are several differences
   d. the hammer Mjolnir. However, there are several differences

5. Which of the following would be the best choice for this sentence (reproduced below)?

(5) <u>For example, Loki, the god of mischief, isn't Thor's brother.</u>

   a. NO CHANGE
   b. For example, Loki the god of mischief isn't Thor's brother.
   c. For example, Loki the god of mischief, isn't Thor's brother.
   d. For example Loki, the god of mischief, isn't Thor's brother.

6. Which of the following would be the best choice for this sentence (reproduced below)?

(6) <u>This is also symbolized in his prized weapon Mjolnir a magic hammer.</u>

   a. NO CHANGE
   b. This is also symbolized in his prized weapon, Mjolnir a magic hammer.
   c. This is also symbolized in his prized weapon, Mjolnir, a magic hammer.
   d. This is also symbolized in his prized weapon Mjolnir, a magic hammer.

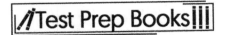

7. Which of the following would be the best choice for this sentence (reproduced below)?

A hammer is both a weapon and a (7) tool, but why would a god favor a seemingly everyday object?

a. NO CHANGE
b. tool; why would a god favor a seemingly everyday object?
c. tool, but, why would a god favor a seemingly everyday object?
d. tool, however, why would a god favor a seemingly everyday object?

8. Which of the following would be the best choice for this sentence (reproduced below)?

The hammer tempers raw iron, (8) ore that is in an chaotic state of impurities and shapelessness, to create an item of worth.

a. NO CHANGE
b. ore that is in a chaotic state of impurities and shapelessness
c. ore that has the impurities and shapelessness of a chaotic state
d. ore that is in an chaotic state, of impurities and shapelessness,

9. Which of the following would be the best choice for this sentence (reproduced below)?

(9) Well the truth is that every culture throughout time, including ours, needs heroes.

a. NO CHANGE
b. Well, the truth is, every culture throughout time, including ours,
c. Well, every culture throughout time, including ours, in truth
d. Well, the truth is that every culture throughout time, including ours,

Questions 10–18 are based on the following passage:

In our essay and class discussion, (10) we came to talking about mirrors. It was an excellent class in which we focused on an article written by Salman Rushdie that compared the homeland to a mirror. (11) Essentially this mirror was an metaphor for us and our homeland. (12) When we look at our reflection we see the culture, our homeland staring back at us. An interesting analogy, but the conversation really began when we read that Rushdie himself stated that the cracked mirror is more valuable than a whole one. But why?

(13) After reflecting on the passage I found the answer to be simple. The analogy reflects the inherent nature of human individuality. The cracks in the mirror represent different aspects of our own being. Perhaps it is our personal views, our hobbies, or our differences with other people, but (14) whatever it is that makes us unique defines us, even while we are part of a big culture. (15) What this tells us is that we can have a homeland, but ultimately we ourselves are each different in it.

Just because one's (16) mirror is cracked, the individuals isn't disowned from the actual, physical homeland and culture within. It means that the homeland is uniquely perceived by the (17) individual beholding it and that there are in fact many aspects to culture itself. Like the various cracks, a culture has religion, language, and many other factors that form to make it whole. What this idea does is invite the viewer to accept their own view of their culture as a whole.

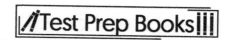

Like in Chandra's *Love and Longing in Bombay*, a single homeland has many stories to tell. Whether one is a cop or a retired war veteran, the individual will perceive the different aspects of the world with unduplicated eyes. (18) Rushdie, seems to be urging his readers to love their culture but to not be pressured by the common crowd. Again, the cracks represent differences which could easily be interpreted as views about the culture, so what this is saying is to accept the culture but accept oneself as well.

From the essay "Portals to Homeland: Mirrors"

10. Which of the following would be the best choice for this sentence (reproduced below)?

In our essay and class discussion, (10) we came to talking about mirrors.

a. NO CHANGE
b. we were talking about
c. we talked about
d. we came to talk about

11. Which of the following would be the best choice for this sentence (reproduced below)?

(11) Essentially this mirror is an metaphor for us and our homeland.

a. NO CHANGE
b. Essentially, this mirror is a metaphor for us and our homeland.
c. Essentially, this mirror is an metaphor for us and our homeland.
d. Essentially this mirror is an metaphor, for us and our homeland.

12. Which of the following would be the best choice for this sentence (reproduced below)?

(12) When we look at our reflection we see the culture, our homeland staring back at us.

a. NO CHANGE
b. When we look at our reflection we see our culture our homeland staring back at us.
c. When we look at our reflection we saw our culture, our homeland, staring back at us.
d. Looking at our reflection we see our culture as our homeland is staring back at us.

13. Which of the following would be the best choice for this sentence (reproduced below)?

(13) After reflecting on the passage I found the answer to be simple.

a. NO CHANGE
b. After reflecting on the passage; I found the answer to be simple.
c. After reflecting on the passage I finding the answer to be simple.
d. After reflecting on the passage, I found the answer to be simple.

14. Which of the following would be the best choice for this sentence (reproduced below)?

Perhaps it is our personal views, our hobbies, or our differences with other people, but (14) <u>whatever it is that makes us unique defines us, even while we are part of a big culture.</u>

a. NO CHANGE
b. whatever it is, that makes us unique, defines us, even while we are part of a big culture.
c. whatever it is that makes us unique also defines us, even while we are part of a bigger culture.
d. whatever it is that makes us unique defines us, even though we are part of a big culture.

15. Which of the following would be the best choice for this sentence (reproduced below)?

(15) <u>What this tells us is that we can have a homeland, but ultimately we ourselves are each different in it.</u>

a. NO CHANGE
b. What this tells us is that we can have a homeland, but ultimately, we ourselves are each different in it.
c. What this tells us is that we can have a homeland, however, ultimately, we ourselves are each different in it.
d. What this tells us is that we can have a homeland, ultimately we ourselves are each different in it.

16. Which of the following would be the best choice for this sentence (reproduced below)?

Just because one's (16) <u>mirror is cracked, the individuals isn't disowned</u> from the actual, physical homeland and culture within.

a. NO CHANGE
b. mirror is cracked, the individuals will not be disowned
c. mirror is cracked, the individuals aren't disowned
d. mirror is cracked, the individual isn't disowned

17. Which of the following would be the best choice for this sentence (reproduced below)?

It means that the homeland is uniquely perceived by the (17) <u>individual beholding it and that there are, in fact, many aspects</u> of culture itself.

a. NO CHANGE
b. individual beholding it; and that there are in fact many aspects
c. individual beholding it and that there is, in fact, many aspects
d. individual beholding it and there's in fact, many aspects

18. Which of the following would be the best choice for this sentence (reproduced below)?

(18) <u>Rushdie, seems to be urging his readers</u> to love their culture but to not be pressured by the common crowd.

a. NO CHANGE
b. Rushdie seemed to be urging his readers
c. Rushdie, seeming to urge his readers,
d. Rushdie seems to be urging his readers

Questions 19–27 are based on the following passage:

The Odyssey reading performed by Odds Bodkin (19) <u>was an especially rewarding experience.</u> Myths continue to help us (20) <u>understood ancient cultures</u> while still helping us connect to real-world lessons through narrative. (21) <u>While myths were not read exactly the way Bodkin performed *The Odyssey*,</u> his performance truly combined ancient and modern artistic styles.

(22) <u>Originally, myths were not written down instead, stories like The Odyssey</u> were transferred to written form thousands of years after they were originally told. *Told* is a (23) <u>key term here: myths were passing on to generations orally.</u> They were sung, or at least accompanied by music. Of course, we no longer gather in halls to hear (24) <u>myths; we read it, or</u> perhaps listen to audio versions. The Odds Bodkin reading was unique in that it offered a traditional glimpse into what it might have been like to receive a myth in its original form of delivery.

What I thought was especially interesting, was that Bodkin used his guitar to build an atmosphere or tempo around the story. In points of action, like when the wooden horse was being moved into Troy, or when Odysseus leads his men in search of the missing crew (lotus episode), he played a signature tune. (25) <u>This creates a kind of soundtrack for the story,</u> providing a unique feeling for the scene. While (26) <u>there is no visual stimuli like in motion picture,</u> one is moved by the guitar's playing. Accompanied by his dramatic reading, the music reflects the action of the plot and even emotions/insights of the characters. Interestingly, Bodkin also used physical gestures to tell the story as well. He occasionally acted out scenes with his hands and sometimes with sound effects.

The best scene was when Bodkin played the Cyclops, Polyphemus, grabbing and gorging himself on Odysseus' men. (27) <u>In addition from bringing a kind of animation</u> to the presentation, it linked the story to the physical world too. He wasn't just reading, he was acting. It gave this a three-dimensional aspect of the telling of *The Odyssey* that I thought was especially great.

19. Which of the following would be the best choice for this sentence (reproduced below)?

The Odyssey reading performed by Odds Bodkin (19) <u>was an especially rewarding experience.</u>

a. NO CHANGE
b. was, an especially, rewarding experience.
c. was a especially rewarding experience.
d. was an especially rewarded experience.

20. Which of the following would be the best choice for this sentence (reproduced below)?

Myths continue to help us (20) <u>understood ancient cultures</u> while still helping us connect to real-world lessons through narrative.

a. NO CHANGE
b. understand ancient cultures
c. understood ancient cultures,
d. understanding ancient cultures

21. Which of the following would be the best choice for this sentence (reproduced below)?

(21) <u>While myths were not read exactly the way Bodkin performed The Odyssey,</u> his performance truly combined ancient and modern artistic styles.

a. NO CHANGE
b. While myths were not exactly the way Bodkin read/performed *The Odyssey*,
c. While myths were not read, exactly, the way Bodkin performed *The Odyssey*,
d. While myths were not read exactly the way Bodkin performed *The Odyssey*.

22. Which of the following would be the best choice for this sentence (reproduced below)?

(22) <u>Originally, myths were not written down instead, stories like The Odyssey</u> were transferred to written form thousands of years after they were originally told.

a. NO CHANGE
b. Myths, originally, were not written down instead, stories like *The Odyssey*
c. Originally myths were not written down; instead stories like *The Odyssey*
d. Originally, myths were not written down. Instead, stories like *The Odyssey*

23. Which of the following would be the best choice for this sentence (reproduced below)?

Told is a (23) <u>key term here: myths were passing on to generations orally.</u>

a. NO CHANGE
b. key term here: myths were passed on to generations orally.
c. key term here: myths passed on to generations orally.
d. key term here: myths pass on to generations orally.

24. Which of the following would be the best choice for this sentence (reproduced below)?

Of course, we no longer gather in halls to hear (24) <u>myths, we read it, or</u> perhaps listen to audio versions.

a. NO CHANGE
b. myths; we read it or
c. myths; they are read, or
d. myths; we read them, or

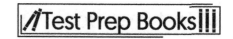

25. Which of the following would be the best choice for this sentence (reproduced below)?

(25) This creates a kind of soundtrack for the story, providing a unique feeling for the scene.

a. NO CHANGE
b. This creates a kind of soundtrack for the story;
c. This creates, a kind of soundtrack, for the story,
d. This creates a kinds of soundtrack for the story,

26. Which of the following would be the best choice for this sentence (reproduced below)?

While (26) there is no visual stimuli like in motion picture, one is moved by the guitar's playing.

a. NO CHANGE
b. there is no visual stimuli, like in motion picture,
c. there are no visual stimuli, like in motion pictures,
d. there are no visual stimuli like in motion picture,

27. Which of the following would be the best choice for this sentence (reproduced below)?

(27) In addition from bringing a kind of animation to the presentation, it linked the story to the physical world too.

a. NO CHANGE
b. In addition to bringing a kind of animation
c. In addition too bringing, a kind of animation,
d. In addition for bringing a kind of animation

Questions 28–36 are based on the following passage:

Quantum mechanics, which describes how the universe works on its smallest scale, is inherently weird. Even the founders of the field (28) including Max Planck, Werner Heisenberg, and Wolfgang Pauli unsettled by the new theory's implications. (29) Instead of a deterministic world where everything can be predicted by equations, events at the quantum scale are purely probabilistic. (30) Every outcome exist simultaneously, while the actual act of observation forces nature to choose one path.

In our everyday lives, (31) this concept of determinism, is actually expressed in the thought experiment of Schrödinger's cat. Devised by Erwin Schrödinger, one of the founders of quantum mechanics, (32) it's purpose is to show how truly strange the framework is. Picture a box containing a cat, a radioactive element, and a vial of poison. (33) If the radioactive element decays, it will release the poison and kill the cat. The box is closed, so there is no way for anyone outside to know what is happening inside. Since the cat's status—alive and dead—are mutually exclusive, only one state can exist. (34) What quantum mechanics says however is that the cat is simultaneously alive and dead, existing in both states until the box's lid is removed and one outcome is chosen.

(35) Further confounding our sense of reality, Louis de Broglie proposed that, on the smallest scales, particles and waves are indistinguishable. This builds on Albert Einstein's famous theory that matter and energy are interchangeable. Although there isn't apparent evidence for this in

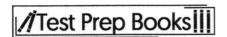

our daily lives, various experiments have shown the validity of quantum mechanics. One of the most famous experiments is the double-slit experiment, which initially proved the wave nature of light. When shone through parallel slits onto a screen, (36) light creates a interference pattern of alternating bands of light and dark. But when electrons were fired at the slits, the act of observation changed the outcome. If observers monitored which slit the electrons travelled through, only one band was seen on the screen. This is expected, since we know electrons act as particles. However, when they monitored the screen only, an interference pattern is created—implying that the electrons behaved as waves!

28. Which of the following would be the best choice for this sentence (reproduced below)?

(28) Even the founders of the field <u>including Max Planck, Werner Heisenberg, and Wolfgang Pauli unsettled by the new theory's implications.</u>

a. NO CHANGE
b. including Max Planck, Werner Heisenberg, and Wolfgang Pauli; unsettled by the new theory's implications.
c. including Max Planck, Werner Heisenberg, and Wolfgang Pauli were unsettled by the new theories' implications.
d. including Max Planck, Werner Heisenberg, and Wolfgang Pauli were unsettled by the new theory's implications.

29. Which of the following would be the best choice for this sentence (reproduced below)?

(29) <u>Instead of a deterministic world where everything can be predicted by equations,</u> events at the quantum scale are purely probabilistic.

a. NO CHANGE
b. Instead, of a deterministic world where everything can be predicted by equations,
c. Instead of a deterministic world where everything can be predicting by equations,
d. Instead of a deterministic world, where everything can be predicted by equations,

30. Which of the following would be the best choice for this sentence (reproduced below)?

(30) <u>Every outcome exist simultaneously,</u> while the actual act of observation forces nature to choose one path.

a. NO CHANGE
b. Each of these outcome exist simultaneously,
c. Every outcome, existing simultaneously,
d. Every outcome exists simultaneously,

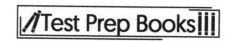

31. Which of the following would be the best choice for this sentence (reproduced below)?

In our everyday lives, (31) this concept of determinism, is actually expressed in the thought experiment of Schrödinger's cat.

a. NO CHANGE
b. this concept of determinism is actually expressed
c. this, concept of determinism, is actually expressed
d. this concept of determinism, is expressed actually

32. Which of the following would be the best choice for this sentence (reproduced below)?

Devised by Erwin Schrödinger, one of the founders of quantum mechanics, (32) it's purpose is to show how truly strange the framework is.

a. NO CHANGE
B. its purposes is to show how truly strange
c. its purpose is to show how truly strange
d. it's purpose, showing how truly strange

33. Which of the following would be the best choice for this sentence (reproduced below)?

(33) If the radioactive element decays, it will release the poison and kill the cat.

a. NO CHANGE
b. If, the radioactive element decays, it will release the poison and kill the cat.
c. If the radioactive element decays. It will release the poison and kill the cat.
d. If the radioactive element decays, releasing the poison and kill the cat.

34. Which of the following would be the best choice for this sentence (reproduced below)?

(34) What quantum mechanics says however is that the cat is simultaneously alive and dead, existing in both states until the box's lid is removed and one outcome is chosen.

a. NO CHANGE
b. What quantum mechanics says however, is
c. What quantum mechanics says. However, is
d. What quantum mechanics says, however, is

ACT Practice Test #2

35. Which of the following would be the best choice for this sentence (reproduced below)?

(35) Further confounding our sense of reality, Louis de Broglie proposed that, on the smallest scales, particles and waves are indistinguishable.

a. NO CHANGE
b. Further confounding our sense of reality Louis de Broglie proposed that on the smallest scales, particles and waves are indistinguishable.
c. Further confounding our sense of reality, Louis de Broglie proposed that on the smallest scales, particles and waves are indistinguishable.
d. Further, confounding our sense of reality, Louis de Broglie proposed that, on the smallest scales, particles and waves are indistinguishable.

36. Which of the following would be the best choice for this sentence (reproduced below)?

When shone through parallel slits onto a screen, (36) light creates a interference pattern of alternating bands of light and dark.

a. NO CHANGE
b. light created an interference
c. lights create a interference
d. light, creating an interference,

Questions 37–45 are based on the following passage:

(37) As the adage goes knowledge is power. Those who are smart and understand the world as it is are the most fit to lead. Intelligence (38) doesn't necessarily require a deep understanding of complex scientific principles. Rather, having the basic knowledge of how the world works, particularly how people go about gaining what they need to survive and thrive, (39) are more important.

Any leadership position, whether on the job or informally, tends to be fraught with politics. (40) Smart leaders will engage in critical thinking allowing them to discern ulterior motives and identify propaganda. Besides catching negative intentions, (41) these practice will serve to highlight the positives in group interactions. Gaining insights into different viewpoints will make leaders more (42) receptive to constructive criticism and ideas from unexpected sources. As with many aspects of being a good and well-rounded person, the seeds for this trait are sown in pre-school. Besides facts and figures, students need to be taught critical thinking skills to survive in a world flooded with subliminal messages and scams. (43) Sadly our current society is plagued by many inconvenient truths that are attacked as lies. Wise leaders should recognize when someone is trying to save the world or merely push a political agenda.

Just as important as the knowledge of how the world works (44) is understood of how humanity operates. Leaders should be able to tactfully make friends and influence people using the doctrines of psychology. People who say whatever comes into their heads without thinking demonstrate a lack of basic diplomatic understanding, not to mention a deficiency of self-control and lack of respect. Breaches of courtesy, whether intentional or otherwise, strain relations and can ruin potential alliances.

While the best leaders are tolerant of other (45) <u>cultural practices and diverse perspectives, those who exhibit disregard</u> and unwarranted contempt for others shouldn't be expected to find favor. Knowledge extends past textbook learning to practical awareness, encompassing skills for a successful life as a decent human being all around. These skills include risk management and creative solutions. Someone who fails to hone these abilities—or neglects to apply their knowledge—will likely be overthrown by those they are supposed to lead.

37. Which of the following would be the best choice for this sentence (reproduced below)?

(37) <u>As the adage goes knowledge is power.</u>

a. NO CHANGE
b. Knowledge is power as the adage goes.
c. As the adage goes, "knowledge is power."
d. As, the adage goes, knowledge is power.

38. Which of the following would be the best choice for this sentence (reproduced below)?

Intelligence (38) <u>doesn't necessarily require</u> a deep understanding of complex scientific principles.

a. NO CHANGE
b. doesn't, necessarily, require
c. doesn't require necessarily
d. aren't necessarily require

39. Which of the following would be the best choice for this sentence (reproduced below)?

Rather, having the basic knowledge of how the world works, particularly how people go about gaining what they need to survive and thrive, (39) <u>are more important.</u>

a. NO CHANGE
b. is more important.
c. are the most important.
d. is most important.

40. Which of the following would be the best choice for this sentence (reproduced below)?

(40) <u>Smart leaders will engage in critical thinking allowing them to discern ulterior motives and identify propaganda.</u>

a. NO CHANGE
b. Smart leaders will engage in critical thinking allowing them to discern ulterior motives, and identify propaganda.
c. Smart leaders will engage in critical thinking, allowing them to discern ulterior motives and identify propaganda.
d. Smart leaders will engage in critical thinking allows them to discern ulterior motives and identify propaganda.

41. Which of the following would be the best choice for this sentence (reproduced below)?

Besides catching negative intentions, (41) these practice will serve to highlight the positives in group interactions.

a. NO CHANGE
b. this practices will serve to highlight
c. these practice serve to highlight
d. this practice will serve to highlight

42. Which of the following would be the best choice for this sentence (reproduced below)?

Gaining insights into different viewpoints will make leaders more (42) receptive to constructive criticism and ideas from unexpected sources.

a. NO CHANGE
b. receptive for constructive criticism and ideas from unexpected sources.
c. receptive from unexpected sources to construct criticism and ideas.
d. receptive, to constructive criticism and ideas, from unexpected sources.

43. Which of the following would be the best choice for this sentence (reproduced below)?

(43) Sadly our current society is plagued by many inconvenient truths that are attacked as lies.

a. NO CHANGE
b. Sadly, our current society is plagued by many inconvenient truths that are attacked as lies.
c. Sadly our current society is plagued by many inconvenient truths, that are attacked as lies.
d. Sadly our current society is plaguing by many inconvenient truths that are attacking as lies.

44. Which of the following would be the best choice for this sentence (reproduced below)?

Just as important as the knowledge of how the world works (44) is understood of how humanity operates.

a. NO CHANGE
b. is to understand of how humanity operates.
c. is understanding of how humanity operates.
d. is an understanding of how humanity operates.

45. Which of the following would be the best choice for this sentence (reproduced below)?

While the best leaders are tolerant of other (45) cultural practices and diverse perspectives, those who exhibit disregard and unwarranted contempt for others shouldn't be expected to find favor.

a. NO CHANGE
b. cultural practices and diverse perspectives. Those who exhibit disregard
c. cultural practices, and diverse perspectives, those who exhibit disregard
d. cultural practices and diverse perspectives, those whom exhibit disregard

Questions 46–54 are based on the following passage:

While the various tales do take on the individual perspectives, (46) <u>Chaucer's internal character retain a unique presence throughout the text.</u> *The Canterbury Tales* provides the ultimate example of a story (47) <u>about stories also being a story containing multiple stories.</u> (48) <u>Chaucer does these by compiling embedded narratives</u> from multiple sources while delivering them through a single, primary voice. From the (49) <u>beginning, it's clear that Chaucer</u> is the source of the text. Yet, Chaucer is more than just a storyteller; he's an honest storyteller. Chaucer provides bold character descriptions of (50) <u>these individual, including their faults.</u> This can be seen when Chaucer sheds light on the nature of the Prioress. He describes that, (51) <u>while her supposed to be a pious nun</u>, she is actually a walking façade. The prioress is noted for counterfeiting courtly behavior so that people think highly of her.

With Chaucer providing such detailed observations, the readers have a more complete picture of what is going on and who the other characters really are. Another example of this is (52) <u>the sarcastic and brutal honest description</u> of the miller: "The millere, that for drunken was al pale." This description of the (53) <u>miller's drunkenness shed's light on</u> why he seems so crazy. It's because he's a drunk. Thus, Chaucer's own narrative gives readers a reliable perspective amongst a host of questionable storytellers. He also provides information on the characters that can help the reader grasp their tales more fully.

One of the signature methods of Chaucer's storytelling is that he doesn't just tell a story, his characters do as well. As a writer, Chaucer actually immerses himself into the collective character discussions. In many ways, Chaucer is a vehicle for the characters to tell their individual stories. Chaucer actually takes on the styles and voice of the characters. A good, eloquent tale like the *Knight's Tale* is written that way because that's how the knight would have told it. (54) <u>Likewise, it's also why several tales are poorly written</u>, have crass delivery, or lack a driven plotline. The *Miller's Tale* is rather crazy and crass because he is drunk while telling it. Chaucer seeks to tell the story in the way his drunken character would have.

46. Which of the following would be the best choice for this sentence (reproduced below)?

While the various tales do take on the individual perspectives, (46) <u>Chaucer's internal character retain a unique presence throughout the text.</u>

    a. NO CHANGE
    b. Chaucer's internal character retains a unique presence throughout the text.
    c. Chaucer's internal character retaining a unique presence throughout the text.
    d. Chaucer's internal character retained a unique presence throughout the text.

47. Which of the following would be the best choice for this sentence (reproduced below)?

The Canterbury Tales provides the ultimate example of a story (47) <u>about stories also being a story containing multiple stories.</u>

    a. NO CHANGE
    b. about stories; not to mention stories within a single story.
    c. about stories, also being a story containing multiple stories.
    d. about stories, as well as being a story containing multiple stories.

48. Which of the following would be the best choice for this sentence (reproduced below)?

(48) <u>Chaucer does these by compiling embedded narratives</u> from multiple sources while delivering them through a single, primary voice.

a. NO CHANGE
b. Chaucer does this, compiling embedded narratives
c. Chaucer does this by compiling embedded narratives
d. Chaucer does these through compiling embedded narratives

49. Which of the following would be the best choice for this sentence (reproduced below)?

From the (49) <u>beginning, it's clear that Chaucer</u> is the source of the text.

a. NO CHANGE
b. beginning. It's clear that Chaucer
c. beginning, Chaucer clearly
d. beginning, it's clear, that Chaucer

50. Which of the following would be the best choice for this sentence (reproduced below)?

Chaucer provides bold character descriptions of (50) <u>these individual, including their faults.</u>

a. NO CHANGE
b. these individuals, including their faults
c. these individual including their faults
d. these individuals; including their faults

51. Which of the following would be the best choice for this sentence (reproduced below)?

He describes that, (51) <u>while her supposed to be a pious nun,</u> she is actually a walking façade.

a. NO CHANGE
b. her supposed nun's piety,
c. while her is supposed to be a pious nun,
d. while she's supposed to be a pious nun,

52. Which of the following would be the best choice for this sentence (reproduced below)?

Another example of this is (52) <u>the sarcastic and brutal honest description of</u> the miller: "The millere, that for drunken was al pale."

a. NO CHANGE
b. the sarcastic, and brutal, honest description of
c. the sarcastic and brutally honest description of
d. the sarcastic, brutal honest description of

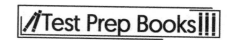

53. Which of the following would be the best choice for this sentence (reproduced below)?

This description of the (53) miller's drunkenness shed's light on why he seems so crazy.

a. NO CHANGE
b. miller's drunkenness sheds light on
c. miller's drunk, which sheds light on
d. miller's drunkenness, shed's light on

54. Which of the following would be the best choice for this sentence (reproduced below)?

Likewise, it's also why several tales are poorly written, have crass delivery, or lack a driven plotline.

a. NO CHANGE
b. Likewise it's also why several tales are poorly written,
c. Likewise it's also why several tales are, poorly written,
d. Likewise, it's also why several tales are, poorly written,

Questions 55–58 are based on the following passage:

(55) Practitioner of Ju-jitsu, and other martial arts, strive away from the act of total violence. We tend not to be violent individuals in the first place because of our training. From the first day on the mat, physical restraint is emphasized. The moves we learn are inherently harmful, (56) so them should used sparingly and not to advance our own satisfaction. The lessons taught in Ju-juitsu urge people to be mindful of their capabilities while also being decisive when defense is required. Having grace, even in the face of danger, is a central theme within martial arts studies.

Everything we do, whether striking a blow or undertaking a certain task, carries responsibility. One must be motivated by pure intention because action cannot be taken back. Life is precious, and its preservation must be top priority. Even when defending against someone who seeks to end your own life, a true Ju-jitsu practitioner should strive to defend themselves while not ending the opponent's life. To abuse this power would only stain the soul and demine the art. (57) Having restraint is testament to how one values life and the possibilities within. Life is about growth, like the study of Ju-jitsu. Ju-jitsu is all about a process of learning and relearning, dedication, struggle, and then understanding. The study (58) of Ju-jitsu, like any trial throughout one's life, is a test of faith and willpower. The end result benefits others as well as oneself.

From the essay "Morality and the Warrior's Path"

55. Which of the following would be the best choice for this sentence (reproduced below)?

(55) Practitioner of Ju-jitsu, and other martial arts, strive away from the act of total violence.

a. NO CHANGE
b. Practitioners of Ju-jitsu, and other martial arts, strive away
c. Practitioners of Ju-jitsu, and other martial arts strive away
d. Practitioner of Ju-jitsu, and other martial arts, strives away

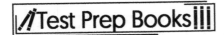

56. Which of the following would be the best choice for this sentence (reproduced below)?

The moves we learn are inherently harmful, (56) <u>so them should used sparingly</u> and not to advance our own satisfaction.

a. NO CHANGE
b. so them should be used sparingly
c. so they should used sparingly
d. so they should be used sparingly

57. Which of the following would be the best choice for this sentence (reproduced below)?

(57) <u>Having restraint is testament to how one values life and the possibilities within.</u>

a. NO CHANGE
b. Having restraint is an testament to how one values life and the possibilities within.
c. Having restraint is a testament to how one values life and the possibilities within.
d. Having restraint is testament to how one values life and the possibilities, within.

58. Which of the following would be the best choice for this sentence (reproduced below)?

The study (58) <u>of Ju-jitsu, like any trial throughout one's life</u> is a test of faith and willpower.

a. NO CHANGE
b. of Ju-jitsu like any trial throughout one's life
c. of Ju-jitsu, like, any trial throughout one's life
d. of Ju-jitsu, like any trial throughout one's life,

Questions 59–67 are based on the following passage:

Our modern society (59) <u>would actually look down on some</u> of Plato's ideas in *The Republic*. But why? (60) <u>Certainly his ideas could help create a more orderly and fair system, but at what cost?</u> The simple truth is that in many of his examples, we see that Plato has taken the individual completely out of the equation. Plato's ideal society is one that places human (61) <u>desire aside to focuses on</u> what will benefit the entire community. To enforce these ideas, Plato seeks to use government to regulate and mandate these rules. This may seem to (62) <u>equalize the population, its possible that this is</u> actually the greatest breech of freedom.

Today, people would think Plato's (63) <u>suggestion to confiscate citizens children</u> and place them in different homes is utterly barbaric. We cannot imagine the pain of losing one's own child to be raised by others. In a modern trial, the judge and (64) <u>jury would see these as kidnapping and</u> ultimately condemn the philosopher. (65) <u>As parents and emotional beings, this seems like cruelty.</u> However, the reason Plato makes this suggestion is to benefit both society and the individual. The reason for this confiscation and placement is for the child to grow to the best of their potential. If the child is brilliant and placed with a (66) <u>similar family than the child will have access to</u> more opportunities. Prosperous or intellectual parents could help develop the child's talents. The child would be placed in an (67) <u>environment that encourage growth.</u> This brilliant child—developed through this different upbringing—would then grow to be a positive presence in the community, a force that would benefit everyone. Thus, in this scenario, Plato *is*

112

thinking of the individual, but resorting to extreme means to *help* them fully develop talents for the good of all the republic.

While our modern views may oppose this idea, we have to acknowledge that this isn't an evil idea. Rather, this is a suggestion to help society through the sacrifice of a few individual liberties. However, one must question how much sacrifice is needed to attain true freedom.

59. Which of the following would be the best choice for this sentence (reproduced below)?

Our modern society (59) <u>would actually look down on some</u> of Plato's ideas in *The Republic*.

    a. NO CHANGE
    b. would actually look down upon some
    c. would actually be looking down on some
    d. would actually look down on something

60. Which of the following would be the best choice for this sentence (reproduced below)?

(60) <u>Certainly his ideas could help create a more orderly and fair system, but at what cost?</u>

    a. NO CHANGE
    b. Certainly, his ideas could help create a more orderly and fair system, at what cost?
    c. Certainly, his ideas could help create a more orderly and fair system, but at what cost?
    d. Certainly his ideas could help create a more orderly and fair system, at what cost?

61. Which of the following would be the best choice for this sentence (reproduced below)?

Plato's ideal society is one that places human (61) <u>desire aside to focuses on</u> what will benefit the entire community.

    a. NO CHANGE
    b. desire aside to focus on
    c. desire aside focusing on
    d. desire aside, focuses on

62. Which of the following would be the best choice for this sentence (reproduced below)?

This may seem to (62) <u>equalize the population, its possible that this is</u> actually the greatest breech of freedom.

    a. NO CHANGE
    b. equalize the population it's possible that this is
    c. equalize the population, but it's possible that this is
    d. equalize the population, however possible that this is

63. Which of the following would be the best choice for this sentence (reproduced below)?

Today, people would think Plato's (63) <u>suggestion to confiscate citizens children</u> and place them in different homes is utterly barbaric.

a. NO CHANGE
b. suggestion for confiscating citizens children
c. suggestion for confiscating citizen's children
d. suggestion to confiscate citizens' children

64. Which of the following would be the best choice for this sentence (reproduced below)?

In a modern trial, the judge and (64) <u>jury would see these as kidnapping and</u> ultimately condemn the philosopher.

a. NO CHANGE
b. jury would see this as kidnapping and
c. jury would see these kidnappings
d. jury would see those as kidnapping and

65. Which of the following would be the best choice for this sentence (reproduced below)?

(65) <u>As parents and emotional beings, this seems like cruelty.</u>

a. NO CHANGE
b. As parents and emotional beings this seems like cruelty.
c. As parents and emotional beings; this seems like cruelty.
d. As parents and emotional beings those seems like cruelty.

66. Which of the following would be the best choice for this sentence (reproduced below)?

If the child is brilliant and placed with a (66) <u>similar family than he will have access to</u> more opportunities.

a. NO CHANGE
b. similar family, than the child will have access to
c. similar family then the child will access to
d. similar family, the child will have access to

67. Which of the following would be the best choice for this sentence (reproduced below)?

The child would be placed in an (67) <u>environment that encourage growth.</u>

a. NO CHANGE
b. environment encouraging growth.
c. environment that encourages growth.
d. environment which encourage growth.

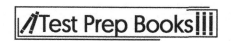

Questions 68–75 are based on the following passage:

Early in my career, (68) <u>a master's teacher shared this thought with me "Education is the last bastion of civility."</u> While I did not completely understand the scope of those words at the time, I have since come to realize the depth, breadth, truth, and significance of what he said. (69) <u>Education provides</u> society with a vehicle for (70) <u>raising it's children to be</u> civil, decent, human beings with something valuable to contribute to the world. It is really what makes us human and what (71) <u>distinguishes</u> us as <u>civilised creatures.</u>

Being "civilized" humans means being "whole" humans. Education must address the mind, body, and soul of students. (72) <u>It would be detrimental to society, only meeting the needs of the mind, if our schools were myopic in their focus.</u> As humans, we are multi-dimensional, multi-faceted beings who need more than head knowledge to survive. (73) <u>The human heart and psyche have to be fed in order for the mind to develop properly, and the body must be maintained and exercised to help fuel the working of the brain. Education is a basic human right, and it allows us to sustain a democratic society in which participation is fundamental to its success. It should inspire students to seek better solutions to world problems and to dream of a more equitable society.</u> Education should never discriminate on any basis, and it should create individuals who are self-sufficient, patriotic, and tolerant of (74) <u>others' ideas.</u>

(75) <u>All children can learn. Although not all children learn in the same manner.</u> All children learn best, however, when their basic physical needs are met, and they feel safe, secure, and loved. Students are much more responsive to a teacher who values them and shows them respect as individual people. Teachers must model at all times the way they expect students to treat them and their peers. If teachers set high expectations for their students, the students will rise to that high level. Teachers must make the well-being of students their primary focus and must not be afraid to let students learn from their own mistakes.

From the essay "Education is Essential to Civilization"

68. Which is the best version of the underlined portion of this sentence (reproduced below)?

Early in my career, (68) <u>a master's teacher shared this thought with me "Education is the last bastion of civility."</u>

a. NO CHANGE
b. a master's teacher shared this thought with me: "Education is the last bastion of civility."
c. a master's teacher shared this thought with me: "Education is the last bastion of civility".
d. a master's teacher shared this thought with me. "Education is the last bastion of civility."

69. Which is the best version of the underlined portion of this sentence (reproduced below)?

(69) <u>Education provides</u> society with a vehicle for raising it's children to be civil, decent, human beings with something valuable to contribute to the world.

a. NO CHANGE
b. Education provide
c. Education will provide
d. Education providing

70. Which is the best version of the underlined portion of this sentence (reproduced below)?

Education provides society with a vehicle for (70) raising it's children to be civil, decent, human beings with something valuable to contribute to the world.

a. NO CHANGE
b. raises its children to be
c. raising its' children to be
d. raising its children to be

71. Which of these, if any, is misspelled?

It is really what makes us human and what (71) distinguishes us as civilised creatures.

a. NO CHANGE
b. distinguishes
c. civilised
d. creatures

72. Which is the best version of the underlined portion of this sentence (reproduced below)?

(72) It would be detrimental to society, only meeting the needs of the mind, if our schools were myopic in their focus.

a. NO CHANGE
b. It would be detrimental to society if our schools were myopic in their focus, only meeting the needs of the mind.
c. Only meeting the needs of our mind, our schools were myopic in their focus, detrimental to society.
d. Myopic is the focus of our schools, being detrimental to society for only meeting the needs of the mind.

73. Which of these sentences, if any, should begin a new paragraph?

(73) The human heart and psyche have to be fed in order for the mind to develop properly, and the body must be maintained and exercised to help fuel the working of the brain. Education is a basic human right, and it allows us to sustain a democratic society in which participation is fundamental to its success. It should inspire students to seek better solutions to world problems and to dream of a more equitable society.

a. NO CHANGE
b. The human heart and psyche have to be fed in order for the mind to develop properly, and the body must be maintained and exercised to help fuel the working of the brain.
c. Education is a basic human right, and it allows us to sustain a democratic society in which participation is fundamental to its success.
d. It should inspire students to seek better solutions to world problems and to dream of a more equitable society.

74. Which is the best version of the underlined portion of this sentence (reproduced below)?

Education should never discriminate on any basis, and it should create individuals who are self-sufficient, patriotic, and tolerant of <u>others' ideas</u>.

a. NO CHANGE
b. other's ideas
c. others ideas
d. others's ideas

75. Which is the best version of the underlined portion of this sentence (reproduced below)?

(75) <u>All children can learn. Although not all children learn in the same manner.</u>

a. NO CHANGE
b. All children can learn although not all children learn in the same manner.
c. All children can learn although, not all children learn in the same manner.
d. All children can learn, although not all children learn in the same manner.

Math

1. Which of the following numbers has the greatest value?
   a. 1.4378
   b. 1.07548
   c. 1.43592
   d. 0.89409
   e. 0.94739

2. The value of 6×12 is the same as:
   a. $2 \times 4 \times 4 \times 2$
   b. $7 \times 4 \times 3$
   c. $6 \times 6 \times 3$
   d. $3 \times 3 \times 4 \times 2$
   e. $5 \times 9 \times 8$

3. This chart indicates how many sales of CDs, vinyl records, and MP3 downloads occurred over the last year. Approximately what percentage of the total sales was from CDs?

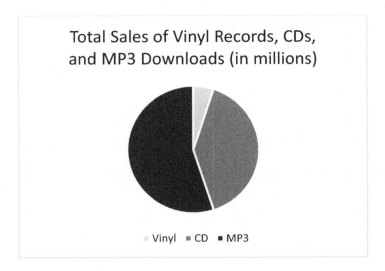

Total Sales of Vinyl Records, CDs, and MP3 Downloads (in millions)

▪ Vinyl ▪ CD ▪ MP3

    a. 55%
    b. 25%
    c. 40%
    d. 5%
    e. 15%

4. Alan currently weighs 200 pounds, but he wants to lose weight to get down to 175 pounds. What is this difference in kilograms? (1 pound is approximately equal to 0.45 kilograms.)
    a. 9 kg
    b. 18.55 kg
    c. 78.75 kg
    d. 90 kg
    e. 11.25 kg

5. Johnny earns $2334.50 from his job each month. He pays $1437 for monthly expenses. Johnny is planning a vacation in 3 months that he estimates will cost $1750 total. How much will Johnny have left over from three months of saving once he pays for his vacation?
    a. $948.50
    b. $584.50
    c. $852.50
    d. $942.50
    e. $742.50

6. Solve the following:

$$\left(\sqrt{36} \times \sqrt{16}\right) - 3^2$$

    a. 30
    b. 21
    c. 15
    d. 13
    e. 25

7. In Jim's school, there are 3 girls for every 2 boys. There are 650 students in total. Using this information, how many students are girls?

    a. 260

    b. 130

    c. 65

    d. 390

    e. 410

8. Kimberley earns $10 an hour babysitting, and after 10 p.m., she earns $12 an hour, with the amount paid being rounded to the nearest hour accordingly. On her last job, she worked from 5:30 p.m. to 11 p.m. In total, how much did Kimberley earn on her last job?

    a. $45

    b. $57

    c. $62

    d. $42

    e. $53

9. Arrange the following numbers from least to greatest value:

$$0.85, \frac{4}{5}, \frac{2}{3}, \frac{91}{100}$$

    a. $0.85, \frac{4}{5}, \frac{2}{3}, \frac{91}{100}$

    b. $\frac{4}{5}, 0.85, \frac{91}{100}, \frac{2}{3}$

    c. $\frac{2}{3}, \frac{4}{5}, 0.85, \frac{91}{100}$

    d. $0.85, \frac{91}{100}, \frac{4}{5}, \frac{2}{3}$

    e. $\frac{4}{5}, \frac{91}{100}, \frac{2}{3}, 0.85$

10. Keith's bakery had 252 customers go through its doors last week. This week, that number increased to 378. Express this increase as a percentage.

    a. 26%

    b. 50%

    c. 35%

    d. 12%

    e. 18%

11. Simplify the following expression:

$$4\frac{2}{3} - 3\frac{4}{9}$$

a. $1\frac{1}{3}$

b. $1\frac{3}{8}$

c. 1

d. $1\frac{2}{3}$

e. $1\frac{2}{9}$

12. Jessica buys 10 cans of paint. Red paint costs $1 per can and blue paint costs $2 per can. In total, she spends $16. How many red cans did she buy?
   a. 2
   b. 3
   c. 4
   d. 5
   e. 6

13. Six people apply to work for Janice's company, but she only needs four workers. How many different groups of four employees can Janice choose?
   a. 6
   b. 10
   c. 15
   d. 36
   e. 42

14. Which of the following is equivalent to the value of the digit 3 in the number 792.134?
   a. 3×10

   b. 3×100

   c. $\frac{3}{10}$

   d. $\frac{3}{100}$

   e. $100 \div 3$

15. In the following expression, which operation should be completed first? $5 \times 6 + (5 + 4) \div 2 - 1$.
   a. Multiplication
   b. Addition
   c. Division
   d. Parentheses
   e. Exponents

16. How will the number 847.89632 be written if rounded to the nearest hundredth?
    a. 847.90
    b. 900
    c. 847.89
    d. 847.896
    e. 850

17. The perimeter of a 6-sided polygon is 56 cm. The length of three of the sides are 9 cm each. The length of two other sides are 8 cm each. What is the length of the missing side?
    a. 11 cm
    b. 12 cm
    c. 13 cm
    d. 10 cm
    e. 9 cm

18. Which of the following is a mixed number?
    a. $16\frac{1}{2}$
    b. 16
    c. $\frac{16}{3}$
    d. $\frac{1}{4}$
    e. $\frac{90}{80}$

19. Change 9.3 to a fraction.
    a. $9\frac{3}{7}$
    b. $\frac{903}{1000}$
    c. $\frac{9.03}{100}$
    d. $3\frac{9}{10}$
    e. $9\frac{3}{10}$

20. What is the value of b in this equation?

$$5b - 4 = 2b + 17$$

    a. 13
    b. 24
    c. 7
    d. 21
    e. 15

21. What is 39% of 164?
    a. 63.96
    b. 23.78
    c. 6,396
    d. 2.38
    e. .0987

22. Katie works at a clothing company and sold 192 shirts over the weekend. One third of the shirts that were sold were patterned, and the rest were solid. Which mathematical expression would calculate the number of solid shirts Katie sold over the weekend?

    a. $192 \times \frac{1}{3}$

    b. $192 \div \frac{1}{3}$

    c. $192 \times (1 - \frac{1}{3})$

    d. $192 \div 3$

    e. $192 \times (1 + \frac{1}{3})$

23. Which four-sided shape is always a rectangle?
    a. Rhombus
    b. Square
    c. Parallelogram
    d. Quadrilateral
    e. Trapezoid

24. A rectangle was formed out of pipe cleaner. Its length was $\frac{1}{2}$ ft, and its width was $\frac{11}{2}$ inches. What is its area in square inches?

    a. $\frac{11}{4}$ inch^2

    b. $\frac{11}{2}$ inch^2

    c. 22 inches^2

    d. $27\frac{1}{3}$ inches^2

    e. 33 inches^2

25. How will $\frac{4}{5}$ be written as a percent?
    a. 40 percent
    b. 125 percent
    c. 90 percent
    d. 80 percent
    e. 25 percent

26. If Danny takes 48 minutes to walk 3 miles, how long should it take him to walk 5 miles maintaining the same speed?
    a. 32 min
    b. 64 min
    c. 80 min
    d. 96 min
    e. 105 min

27. A solution needs 5 ml of saline for every 8 ml of medicine given. How much saline is needed for 45 ml of medicine?
    a. $\frac{225}{8}$ ml
    b. 72 ml
    c. 28 ml
    d. $\frac{45}{8}$ ml
    e. 84 ml

28. What unit of volume is used to describe the following 3-dimensional shape?

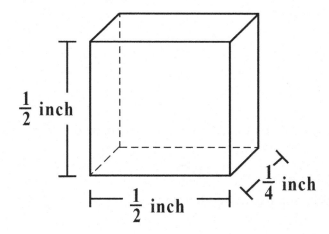

    a. Square inches
    b. Inches
    c. Cubic inches
    d. Squares
    e. Cubic feet

29. Which common denominator would be used in order to evaluate $\frac{2}{3} + \frac{4}{5}$?
    a. 15
    b. 3
    c. 5
    d. 10
    e. 12

30. The diameter of a circle measures 5.75 centimeters. What tool could be used to draw such a circle?
    a. Ruler
    b. Meter stick
    c. Compass
    d. Yard stick
    e. Protractor

31. A piggy bank contains 12 dollars' worth of nickels. A nickel weighs 5 grams, and the empty piggy bank weighs 1050 grams. What is the total weight of the full piggy bank?
    a. 1,110 grams
    b. 1,200 grams
    c. 2,150 grams
    d. 2,250 grams
    e. 2,500 grams

32. Last year, the New York City area received approximately $27\frac{3}{4}$ inches of snow. The Denver area received approximately 3 times as much snow as New York City. How much snow fell in Denver?
    a. $71\frac{3}{4}$ inches

    b. $27\frac{1}{4}$ inches

    c. $89\frac{1}{4}$ inches

    d. $83\frac{1}{4}$ inches

    e. $64\frac{3}{4}$ inches

33. Which of the following would be an instance in which ordinal numbers are used?
    a. Katie scored a 9 out of 10 on her quiz.
    b. Matthew finished second in the spelling bee.
    c. Jacob missed one day of school last month.
    d. Kim was 5 minutes late to school this morning.
    e. Katrina lost 10 lbs. over the summer.

34. How will the following number be written in standard form: $(1 \times 10^4) + (3 \times 10^3) + (7 \times 10^1) + (8 \times 10^0)$
    a. 137
    b. 13,780
    c. 1,378
    d. 13,078
    e. 3,780

35. What is the area of the regular hexagon shown below?

10.39

12

a. 72
b. 124.68
c. 374.04
d. 748.08
e. 676.79

36. The area of a given rectangle is 24 square centimeters. If the measure of each side is multiplied by 3, what is the area of the new figure?
a. 48 cm^2
b. 72 cm^2
c. 111 cm^2
d. 13,824 cm^2
e. 216 cm^2

37. Which of the following is the definition of a prime number?
a. A number that factors only into itself and 1
b. A number greater than one that factors only into itself and 1
c. A number less than 10
d. A number divisible by 10
e. Any number that's greater than 1

38. Add and express in reduced form $\frac{5}{12} + \frac{4}{9}$
a. $\frac{9}{17}$
b. $\frac{1}{3}$
c. $\frac{31}{36}$
d. $\frac{3}{5}$
e. $\frac{15}{29}$

39. Which of the following is the correct order of operations that could be used on a difficult math problem that contained grouping symbols?
a. Parentheses, Exponents, Multiplication, Division, Addition, Subtraction
b. Exponents, Parentheses, Multiplication, Division, Addition, Subtraction
c. Parentheses, Exponents, Addition, Multiplication, Division, Subtraction
d. Parentheses, Exponents, Division, Addition, Subtraction, Multiplication
e. Multiplication, Parentheses, Exponents, Division, Subtraction, Addition

40. Convert $\frac{5}{8}$ to a decimal.
    a. 0.62
    b. 1.05
    c. 0.63
    d. 1.60
    e. 1.25

41. Subtract $9,576 - 891$.
    a. 10,467
    b. 9,685
    c. 8,325
    d. 8,685
    e. 7,564

42. If a teacher was showing a class how to round 245.2678 to the nearest thousandth, which place value would be used to decide whether to round up or round down?
    a. Ten-thousandth
    b. Thousandth
    c. Hundredth
    d. Thousand
    e. Hundred

43. Subtract $50.888 - 13.091$.
    a. 63.799
    b. 63.979
    c. 37.979
    d. 37.797
    e. 36.697

44. Students should line up decimal places within the given numbers before performing which of the following?
    a. Multiplication
    b. Division
    c. Subtraction
    d. Exponents
    e. Parentheses

45. Subtract and express in reduced form $\frac{23}{24} - \frac{1}{6}$.
    a. $\frac{22}{18}$
    b. $\frac{11}{9}$
    c. $\frac{19}{24}$
    d. $\frac{4}{5}$
    e. $\frac{1}{2}$

46. Subtract and express in reduced form $\frac{43}{45} - \frac{11}{15}$.

    a. $\frac{10}{45}$

    b. $\frac{16}{15}$

    c. $\frac{32}{30}$

    d. $\frac{5}{9}$

    e. $\frac{2}{9}$

47. Change 0.56 to a fraction.

    a. $\frac{5.6}{100}$

    b. $\frac{14}{25}$

    c. $\frac{56}{1000}$

    d. $\frac{56}{10}$

    e. $\frac{100}{56}$

48. Multiply $13,114 \times 191$.

    a. 2,504,774

    b. 250,477

    c. 150,474

    d. 2,514,774

    e. 1,759,492

49. Marty wishes to save $150 over a 4-day period. How much must Marty save each day on average?

    a. $37.50

    b. $35

    c. $45.50

    d. $41

    e. $43.50

50. A teacher cuts a pie into 6 equal pieces and takes one away. What topic would she be introducing to the class by using such a visual?

    a. Decimals

    b. Addition

    c. Subtraction

    d. Fractions

    e. Exponents

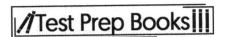

51. Multiply and reduce $\frac{15}{23} \times \frac{54}{127}$.

    a. $\frac{810}{2,921}$

    b. $\frac{81}{292}$

    c. $\frac{69}{150}$

    d. $\frac{810}{2929}$

    e. $\frac{2921}{810}$

52. Which of the following represent one hundred eighty-two billion, thirty-six thousand, four hundred twenty-one and three hundred fifty-six thousandths?

    a. 182,036,421.356
    b. 182,036,421.0356
    c. 182,000,036,421.0356
    d. 182,000,036,421.356
    e. 182,360,000,421.356

53. Divide, express with a remainder $188 \div 16$.

    a. $1\frac{3}{4}$

    b. $111\frac{3}{4}$

    c. $10\frac{3}{4}$

    d. $11\frac{3}{4}$

    e. $3\frac{4}{11}$

54. What other operation could be utilized to teach the process of dividing 9453 by 24 besides division?
    a. Multiplication
    b. Addition
    c. Exponents
    d. Subtraction
    e. Parentheses

55. Bernard can make $80 per day. If he needs to make $300 and only works full days, how many days will this take?
    a. 6
    b. 3
    c. 5
    d. 2
    e. 4

56. A couple buys a house for \$150,000. They sell it for \$165,000. By what percentage did the house's value increase?
    a. 18%
    b. 13%
    c. 15%
    d. 12%
    e. 10%

57. What operation is repeated to evaluate an expression involving an exponent?
    a. Addition
    b. Multiplication
    c. Division
    d. Subtraction
    e. Parentheses

58. Which of the following formulas would correctly calculate the perimeter of a legal-sized piece of paper that is 14 inches long and $8\frac{1}{2}$ inches wide?
    a. $P = 14 + 8\frac{1}{2}$

    b. $P = 14 + 8\frac{1}{2} + 14 + 8\frac{1}{2}$

    c. $P = 14 \times 8\frac{1}{2}$

    d. $P = 14 \times \frac{17}{2}$

    e. $P = 14 - \frac{17}{2}$

59. Which of the following are units that would be taught in a lecture covering the metric system?
    a. Inches, feet, miles, pounds
    b. Millimeters, centimeters, meters, pounds
    c. Kilograms, grams, kilometers, meters
    d. Teaspoons, tablespoons, ounces
    e. Stone, quart, gallon, foot, yard

60. Which important mathematical property is shown in the expression: $(7 \times 3) \times 2 = 7 \times (3 \times 2)$?
    a. Distributive property
    b. Commutative property
    c. Additive inverse
    d. Associative property
    e. Multiplicative Identity property

Reading

Fiction

Questions 1–10 are based on the following passage:

As long ago as 1860 it was the proper thing to be born at home. At present, so I am told, the high gods of medicine have decreed that the first cries of the young shall be uttered upon the anesthetic air of a hospital, preferably a fashionable one. So young Mr. and Mrs. Roger Button were fifty years ahead of style when they decided, one day in the summer of 1860, that their first baby should be born in a hospital. Whether this anachronism had any bearing upon the astonishing history I am about to set down will never be known.

I shall tell you what occurred, and let you judge for yourself.

The Roger Buttons held an enviable position, both social and financial, in ante-bellum Baltimore. They were related to the This Family and the That Family, which, as every Southerner knew, entitled them to membership in that enormous peerage which largely populated the Confederacy. This was their first experience with the charming old custom of having babies— Mr. Button was naturally nervous. He hoped it would be a boy so that he could be sent to Yale College in Connecticut, at which institution Mr. Button himself had been known for four years by the somewhat obvious nickname of "Cuff."

On the September morning <u>consecrated</u> to the enormous event he arose nervously at six o'clock, dressed himself, adjusted an impeccable stock, and hurried forth through the streets of Baltimore to the hospital, to determine whether the darkness of the night had borne in new life upon its bosom.

When he was approximately a hundred yards from the Maryland Private Hospital for Ladies and Gentlemen he saw Doctor Keene, the family physician, descending the front steps, rubbing his hands together with a washing movement—as all doctors are required to do by the unwritten ethics of their profession.

Mr. Roger Button, the president of Roger Button & Co., Wholesale Hardware, began to run toward Doctor Keene with much less dignity than was expected from a Southern gentleman of that picturesque period. "Doctor Keene!" he called. "Oh, Doctor Keene!"

The doctor heard him, faced around, and stood waiting, a curious expression settling on his harsh, medicinal face as Mr. Button drew near.

"What happened?" demanded Mr. Button, as he came up in a gasping rush. "What was it? How is she? A boy? Who is it? What—"

"Talk sense!" said Doctor Keene sharply. He appeared somewhat irritated.

"Is the child born?" begged Mr. Button.

Doctor Keene frowned. "Why, yes, I suppose so—after a fashion." Again he threw a curious glance at Mr. Button.

From *The Curious Case of Benjamin Button* by F.S. Fitzgerald, 1922.

1. According to the passage, what major event is about to happen in this story?
   a. Mr. Button is about to go to a funeral.
   b. Mr. Button's wife is about to have a baby.
   c. Mr. Button is getting ready to go to the doctor's office.
   d. Mr. Button is about to go shopping for new clothes.

2. What kind of tone does the above passage have?
   a. Nervous and Excited
   b. Sad and Angry
   c. Shameful and Confused
   d. Grateful and Joyous

3. As it is used in the fourth paragraph, the word *consecrated* most nearly means:
   a. Numbed
   b. Chained
   c. Dedicated
   d. Moved

4. What does the author mean to do by adding the following statement?

"rubbing his hands together with a washing movement—as all doctors are required to do by the unwritten ethics of their profession."

   a. Suggesting that Mr. Button is tired of the doctor.
   b. Trying to explain the detail of the doctor's profession.
   c. Hinting to readers that the doctor is an unethical man.
   d. Giving readers a visual picture of what the doctor is doing.

5. Which of the following best describes the development of this passage?
   a. It starts in the middle of a narrative in order to transition smoothly to a conclusion.
   b. It is a chronological narrative from beginning to end.
   c. The sequence of events is backwards—we go from future events to past events.
   d. To introduce the setting of the story and its characters.

6. Which of the following is an example of an imperative sentence?
   a. "Oh, Doctor Keene!"
   b. "Talk sense!"
   c. "Is the child born?"
   d. "Why, yes, I suppose so—"

7. As it is used in the first paragraph, the word *anachronism* most nearly means:
   a. Comparison
   b. Misplacement
   c. Aberration
   d. Amelioration

8. This passage can best be described as what type of text?
    a. Expository
    b. Descriptive
    c. Narrative
    d. Persuasive

9. The main purpose of the first paragraph is:
    a. To explain the setting of the narrative and give information about the story.
    b. To present the thesis so that the audience can determine which points are valid later in the text.
    c. To introduce a counterargument so that the author can refute it in the next paragraph.
    d. To provide a description of the speaker's city and the building in which he works.

10. The end of the passage implies to the audience that:
    a. There is bad weather coming.
    b. The doctor thinks Mr. Button is annoying.
    c. The baby and the mother did not make it through labor.
    d. Something is unusual about the birth of the baby.

Social Science

Questions 11–20 are based on the following passage:

I heartily accept the motto, "that government is best which governs least," and I should like to see it acted up to more rapidly and systematically. Carried out, it finally amounts to this, which also I believe—"that government is best which governs not at all," and when men are prepared for it, that will be the kind of government which they will have. Government is at best but an expedient; but most governments are usually, and all governments are sometimes, inexpedient. The objections which have been brought against a standing army, and they are many and weighty, and deserve to prevail, may also at last be brought against a standing government. The standing army is only an arm of the standing government. The government itself, which is only the mode which the people have chosen to execute their will, is equally liable to be abused and perverted before the people can act through it. Witness the present Mexican war, the work of comparatively a few individuals using the standing government as their tool; for, in the outset, the people would not have consented to this measure.

This American government—what is it but a tradition, though a recent one, endeavoring to transmit itself unimpaired to posterity, but each instant losing some of its integrity? It has not the vitality and force of a single living man; for a single man can bend it to his will. It is a sort of wooden gun to the people themselves. But it is not the less necessary for this; for the people must have some complicated machinery or other, and hear its din, to satisfy that idea of government which they have. Governments show thus how successfully men can be imposed on, even impose on themselves, for their own advantage. It is excellent, we must all allow. Yet this government never of itself furthered any enterprise, but by the alacrity with which it got out of its way. It does not keep the country free. It does not settle the West. It does not educate. The character inherent in the American people has done all that has been accomplished; and it would have done somewhat more, if the government had not sometimes got in its way. For government is an expedient by which men would fain succeed in letting one another alone; and, as has been said, when it is most expedient, the governed are most let alone by it. Trade and commerce, if they were not made of India-rubber, would never manage to bounce over the

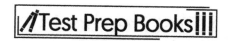

obstacles which legislators are continually putting in their way; and, if one were to judge these men wholly by the effects of their actions and not partly by their intentions, they would deserve to be classed and punished with those mischievous persons who put obstructions on the railroads.

But, to speak practically and as a citizen, unlike those who call themselves no-government men, I ask for, not at once no government, but at once a better government. Let every man make known what kind of government would command his respect, and that will be one step toward obtaining it.

Excerpt from Civil Disobedience by Henry David Thoreau

11. Which phrase best encapsulates Thoreau's use of the term *expedient* in the first paragraph?
    a. A dead end
    b. A state of order
    c. A means to an end
    d. Rushed construction

12. Which best describes Thoreau's view on the Mexican War?
    a. Government is inherently corrupt because it must wage war.
    b. Government can easily be manipulated by a few individuals for their own agenda.
    c. Government is a tool for the people, but it can also act against their interest.
    d. The Mexican War was a necessary action, but not all the people believed this.

13. What is Thoreau's purpose for writing?
    a. His goal is to illustrate how government can function if ideals are maintained.
    b. He wants to prove that true democracy is the best government, but it can be corrupted easily.
    c. Thoreau reflects on the stages of government abuses.
    d. He is seeking to prove that government is easily corruptible and inherently restrictive of individual freedoms that can simultaneously affect the whole state.

14. Which example best supports Thoreau's argument?
    a. A vote carries in the Senate to create a new road tax.
    b. The president vetoes the new FARM bill.
    c. Prohibition is passed to outlaw alcohol.
    d. Trade is opened between the United States and Iceland.

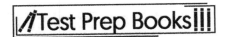

15. Which best summarizes this section from the following passage?

"This American government—what is it but a tradition, though a recent one, endeavoring to transmit itself unimpaired to posterity, but each instant losing some of its integrity? It has not the vitality and force of a single living man; for a single man can bend it to his will. It is a sort of wooden gun to the people themselves."

a. The government may be instituted to ensure the protections of freedoms, but this is weakened by the fact that it is easily manipulated by individuals.
b. Unlike an individual, government is uncaring.
c. Unlike an individual, government has no will, making it more prone to be used as a weapon against the people.
d. American government is modeled after other traditions but actually has greater potential to be used to control people.

16. According to Thoreau, what's the main reason why government eventually fails to achieve progress?
a. There are too many rules.
b. Legislation eventually becomes a hindrance to the lives and work of everyday people.
c. Trade and wealth eventually become the driving factor of those in government.
d. Government doesn't separate religion and state.

17. What type of passage is this?
a. Narrative
b. Descriptive
c. Persuasive
d. Expository

18. As it is used in the first paragraph, the word *liable* most nearly means:
a. Paramount
b. Inconceivable
c. Susceptible
d. Detrimental

19. According to the passage, which government is Thoreau talking about?
a. Mexican
b. American
c. Chinese
d. British

20. As it is used in the second paragraph, the word *posterity* most nearly means:
a. Persons of royal lineage.
b. All future generations of people.
c. A person involved in directing education.
d. A person who offers views on important life questions

Humanities

Questions 21–30 are based on the following passage:

Four hundred years ago, in 1612, the north-west of England was the scene of England's biggest peacetime witch trial: the trial of the Lancashire witches. Twenty people, mostly from the Pendle area of Lancashire, were imprisoned in the castle as witches. Ten were hanged, one died in gaol, one was sentenced to stand in the pillory, and eight were acquitted. The 2012 anniversary sees a small flood of commemorative events, including works of fiction by Blake Morrison, Carol Ann Duffy, and Jeanette Winterson. How did this witch trial come about, and what accounts for its enduring fame?

We know so much about the Lancashire Witches because the trial was recorded in unique detail by the clerk of the court, Thomas Potts, who published his account soon afterwards as *The Wonderful Discovery of Witches in the County of Lancaster.* I have recently published a modern-English edition of this book, together with an essay piecing together what we know of the events of 1612. It has been a fascinating exercise, revealing how Potts carefully edited the evidence, and also how the case against the "witches" was constructed and manipulated to bring about a spectacular show trial. It all began in mid-March when a pedlar from Halifax named John Law had a frightening encounter with a poor young woman, Alizon Device, in a field near Colne. He refused her request for pins and there was a brief argument during which he was seized by a fit that left him with "his head … drawn awry, his eyes and face deformed, his speech not well to be understood; his thighs and legs stark lame." We can now recognize this as a stroke, perhaps triggered by the stressful encounter. Alizon Device was sent for and surprised all by confessing to the bewitching of John Law and then begged for forgiveness.

When Alizon Device was unable to cure the pedlar, the local magistrate, Roger Nowell was called in. Characterized by Thomas Potts as "God's justice" he was alert to instances of witchcraft, which were regarded by the Lancashire's puritan-inclined authorities as part of the cultural rubble of "popery"—Roman Catholicism—long overdue to be swept away at the end of the country's very slow protestant reformation. "With weeping tears" Alizon explained that she had been led astray by her grandmother, "old Demdike," well-known in the district for her knowledge of old Catholic prayers, charms, cures, magic, and curses. Nowell quickly interviewed Alizon's grandmother and mother, as well as Demdike's supposed rival, "old Chattox" and her daughter Anne. Their panicky attempts to explain themselves and shift the blame to others eventually only ended up incriminating them, and the four were sent to Lancaster gaol in early April to await trial at the summer assizes. The initial picture revealed was of a couple of poor, marginal local families in the forest of Pendle with a longstanding reputation for magical powers, which they had occasionally used at the request of their wealthier neighbours. There had been disputes but none of these were part of ordinary village life. Not until 1612 did any of this come to the attention of the authorities.

The net was widened still further at the end of April when Alizon's younger brother James and younger sister Jennet, only nine years old, came up between them with a story about a "great meeting of witches" at their grandmother's house, known as Malkin Tower. This meeting was presumably to discuss the plight of those arrested and the threat of further arrests, but according to the evidence extracted from the children by the magistrates, a plot was hatched to blow up Lancaster castle with gunpowder, kill the gaoler, and rescue the imprisoned witches. It was, in short, a conspiracy against royal authority to rival the gunpowder plot of 1605—

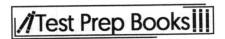

something to be expected in a county known for its particularly strong underground Roman Catholic presence.

Those present at the meeting were mostly family members and neighbours, but they also included Alice Nutter, described by Potts as "a rich woman [who] had a great estate, and children of good hope: in the common opinion of the world, of good temper, free from envy or malice." Her part in the affair remains mysterious, but she seems to have had Catholic family connections, and may have been one herself, providing an added motive for her to be prosecuted.

"The Lancashire Witches 1612–2012" by Robert Poole.

This article (The Lancaster Witches 1612-2012) was originally published in The Public Domain Review under a Creative Commons Attribution-ShareAlike 3.0. If you wish to reuse it, please see: http://publicdomainreview.org/legal/

21. What's the point of this passage, and why did the author write it?
 a. The author is documenting a historic witchcraft trial while uncovering/investigating the role of suspicion and anti-Catholicism in the events.
 b. The author seeks long-overdue reparations for the ancestors of those accused and executed for witchcraft in Lancashire.
 c. The author is educating the reader about actual occult practices of the 1600s.
 d. The author argues that the Lancashire witch trials were more brutal than the infamous Salem trials.

22. Which term best captures the meaning of the author's use of *enduring* in the first paragraph?
 a. Un-original
 b. Popular
 c. Wicked
 d. Circumstantial

23. What textual information is present within the passage that most lends itself to the author's credibility?
 a. His prose is consistent with the time.
 b. This is a reflective passage; the author doesn't need to establish credibility.
 c. The author cites specific quotes.
 d. The author has published a modern account of the case and has written on the subject before.

24. What might the following excerpt suggest about the trial or, at the very least, Thomas Potts' account of the trial(s)?

"It has been a fascinating exercise, revealing how Potts carefully edited the evidence, and also how the case against the 'witches' was constructed and manipulated to bring about a spectacular show trial."

 a. The events were so grand that the public was allowed access to such a spectacular set of cases.
 b. Sections may have been exaggerated or stretched to create notoriety on an extraordinary case.
 c. Evidence was faked, making the trial a total farce.
 d. The trial was corrupt from the beginning.

25. Which statement best describes the political atmosphere of the 1600s that influenced the Alizon Device witch trial/case?

    a. Fear of witches was prevalent during this period.

    b. Magistrates were seeking ways to cement their power during this period of unrest.

    c. In a highly superstitious culture, the Protestant church and government were highly motivated to root out any potential sources that could undermine the current regime.

    d. Lancashire was originally a prominent area for pagan celebration, making the modern Protestants very weary of whispers of witchcraft and open to witch trials to resolve any potential threats to Christianity.

26. Which best describes the strongest "evidence" used in the case against Alizon and the witches?

    a. Knowledge of the occult and witchcraft

    b. "Spectral evidence"

    c. Popular rumors of witchcraft and Catholic association

    d. Self-incriminating speech

27. What type of passage is this?

    a. Persuasive

    b. Expository

    c. Narrative

    d. Descriptive

28. According to the passage, how many people were arrested as witches in the Lancashire trials?

    a. 10

    b. 20

    c. 30

    d. 40

29. As it is used in the first paragraph, the word *commemorative* most nearly means:

    a. Associated with being acquitted

    b. An act of disloyalty

    c. A circumstance to be disliked

    d. In honor of something

30. According to the passage, what is Malkin tower?

    a. The building where the trial of the Lancashire witches took place.

    b. The grandmother's house of the peddler who sought revenge on Alizon.

    c. Alizon's grandmother's house where a meeting of witches was held.

    d. The residence of a jury member who witnessed the cursing of the peddler.

Natural Science

Questions 31–40 are based upon the following passage:

Insects as a whole are preeminently creatures of the land and the air. This is shown not only by the possession of wings by a vast majority of the class, but by the mode of breathing to which reference has already been made, a system of branching air-tubes carrying atmospheric air with its combustion-supporting oxygen to all the insect's tissues. The air gains access to these tubes through a number of paired air-holes or spiracles, arranged segmentally in series.

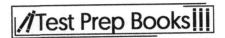

It is of great interest to find that, nevertheless, a number of insects spend much of their time under water. This is true of not a few in the perfect winged state, as for example aquatic beetles and water-bugs ('boatmen' and 'scorpions') which have some way of protecting their spiracles when submerged, and, possessing usually the power of flight, can pass on occasion from pond or stream to upper air. But it is advisable in connection with our present subject to dwell especially on some insects that remain continually under water till they are ready to undergo their final moult and attain the winged state, which they pass entirely in the air. The preparatory instars of such insects are aquatic; the adult instar is aerial. All may-flies, dragon-flies, and caddis-flies, many beetles and two-winged flies, and a few moths thus divide their life-story between the water and the air. For the present we confine attention to the Stone-flies, the May-flies, and the Dragon-flies, three well-known orders of insects respectively called by systematists the Plecoptera, the Ephemeroptera and the Odonata.

In the case of many insects that have aquatic larvae, the latter are provided with some arrangement for enabling them to reach atmospheric air through the surface-film of the water. But the larva of a stone-fly, a dragon-fly, or a may-fly is adapted more completely than these for aquatic life; it can, by means of gills of some kind, breathe the air dissolved in water.

This excerpt is from *The Life-Story of Insects* by Geo H. Carpenter

31. Which statement best details the central idea in this passage?
    a. It introduces certain insects that transition from water to air.
    b. It delves into entomology, especially where gills are concerned.
    c. It defines what constitutes as insects' breathing.
    d. It invites readers to have a hand in the preservation of insects.

32. Which definition most closely relates to the usage of the word *moult* in the passage?
    a. An adventure of sorts, especially underwater
    b. Mating act between two insects
    c. The act of shedding part or all of the outer shell
    d. Death of an organism that ends in a revival of life

33. What is the purpose of the first paragraph in relation to the second paragraph?
    a. The first paragraph serves as a cause, and the second paragraph serves as an effect.
    b. The first paragraph serves as a contrast to the second.
    c. The first paragraph is a description for the argument in the second paragraph.
    d. The first and second paragraphs are merely presented in a sequence.

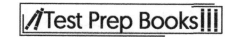

34. What does the following sentence most nearly mean?

The preparatory instars of such insects are aquatic; the adult instar is aerial.

a. The volume of water is necessary to prep the insect for transition rather than the volume of the air.
b. The abdomen of the insect is designed like a star in the water as well as the air.
c. The stage of preparation in between molting is acted out in the water, while the last stage is in the air.
d. These insects breathe first in the water through gills yet continue to use the same organs to breathe in the air.

35. Which of the statements reflect information that one could reasonably infer based on the author's tone?
a. The author's tone is persuasive and attempts to call the audience to action.
b. The author's tone is passionate due to excitement over the subject and personal narrative.
c. The author's tone is informative and exhibits interest in the subject of the study.
d. The author's tone is somber, depicting some anger at the state of insect larvae.

36. Which statement best describes stoneflies, mayflies, and dragonflies?
a. They are creatures of the land and the air.
b. They have a way of protecting their spiracles when submerged.
c. Their larvae can breathe the air dissolved in water through gills of some kind.
d. The preparatory instars of these insects are aerial.

37. According to the passage, what is true of "boatmen" and "scorpions"?
a. They have no way of protecting their spiracles when submerged.
b. They have some way of protecting their spiracles when submerged.
c. They usually do not possess the power of flight.
d. They remain continually under water till they are ready to undergo their final moult.

38. The last paragraph indicates that the author believes
a. That the stonefly, dragonfly, and mayfly larvae are better prepared to live beneath the water because they have gills that allow them to do so.
b. That the stonefly is different from the mayfly because the stonefly can breathe underwater and the mayfly can only breathe above water.
c. That the dragonfly is a unique species in that its larvae lives mostly underwater for most of its young life.
d. That the stonefly larvae can breathe only by reaching the surface film of the water.

39. According to the passage, why are insects as a whole preeminently creatures of the land and the air?
a. Because insects are born on land but eventually end up adapting to life underwater for the rest of their adult lives.
b. Because most insects have legs made for walking on land and tube-like structures on their bellies for skimming the water.
c. Because a vast majority of insects have wings and also have the ability to breathe underwater.
d. Because most insects have a propulsion method specifically designed for underwater use, but they can only breathe on land.

40. As it is used in the first paragraph, the word *preeminently* most nearly means:
    a. Unknowingly
    b. Above all
    c. Most truthfully
    d. Not importantly

Science

Passage 1

Questions 1–5 pertain to Passage 1:

Predators are animals that eat other animals. Prey are animals that are eaten by a predator. Predators and prey have a distinct relationship. Predators rely on the prey population for food and nutrition. They evolve physically to catch their prey. For example, they develop a keen sense of sight, smell, or hearing. They may also be able to run very fast or camouflage to their environment in order to sneak up on their prey. Likewise, the prey population may develop these features to escape and hide from their predators. As predators catch more prey, the prey population dwindles. With fewer prey to catch, the predator population also dwindles. This happens in a cyclical manner over time.

Figure 1 below shows the cyclical population growth in a predator-prey relationship.

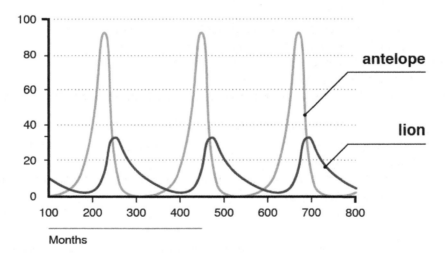

Figure 2 below shows a predator-prey cycle in a circular picture diagram

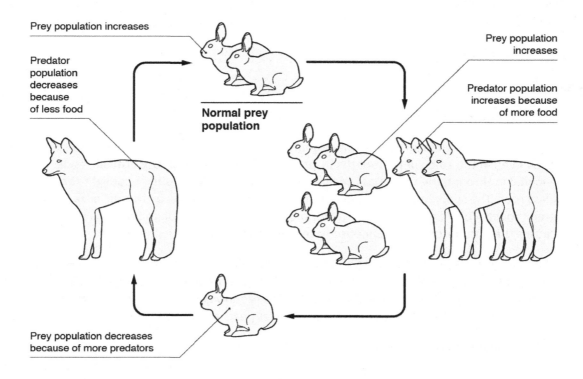

1. Looking at Figure 1, approximately how long is one cycle of the prey population, which includes the population being low, reaching a peak, and then becoming low again?
   a. 300 months
   b. 200 months
   c. 800 months
   d. 400 months

2. In Figure 2, which animal is the predator?
   a. Both the fox and rabbit
   b. Fox only
   c. Rabbit only
   d. Neither the fox nor the rabbit

3. What causes the predator population to decrease?
   a. When there's an increase in the prey population
   b. When winter arrives
   c. When the prey start attacking the predators
   d. When there are fewer prey to find

4. What causes the prey population to increase?
   a. When the predator population decreases, so more prey survive and reproduce.
   b. When there's an increase in the predator population
   c. When there's more sunlight
   d. The prey population always remains the same size.

5. Which is NOT a feature that a prey population can develop to hide from their predator?
   a. Keen sense of smell
   b. Camouflage ability
   c. A loud voice
   d. Keen sense of hearing

Passage 2

Questions 6–10 pertain to Passage 2:

Greenhouses are glass structures that people grow plants in. They allow plants to survive and grow even in the cold winter months by providing light and trapping warm air inside. Light is allowed in through the clear glass walls and roof. Warm air comes in as sunlight through the glass roof. The sunlight is converted into heat, or infrared energy, by the surfaces inside the greenhouse. This heat energy then takes longer to pass back through the glass surfaces and causes the interior of the greenhouse to feel warmer than the outside climate.

Plants may grow better inside a greenhouse versus outside for several reasons. There is more control of the temperature and humidity of the environment inside the greenhouse. The carbon dioxide produced by plants is trapped inside the greenhouse and can increase the rate of photosynthesis of the plants. There are also fewer pests and diseases inside the greenhouse.

Figure 1 below shows how a greenhouse works.

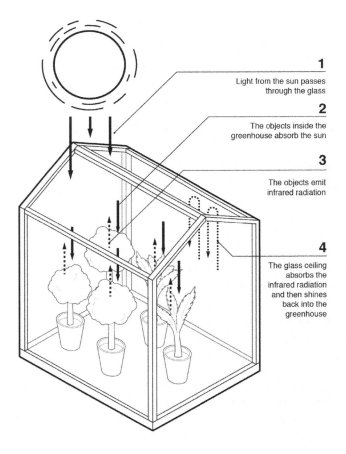

1
Light from the sun passes through the glass

2
The objects inside the greenhouse absorb the sun

3
The objects emit infrared radiation

4
The glass ceiling absorbs the infrared radiation and then shines back into the greenhouse

Scientist A wants to compare how a tomato plant grows inside a greenhouse versus outside a greenhouse.

Figure 2 below shows a graph of her results over 3 months.

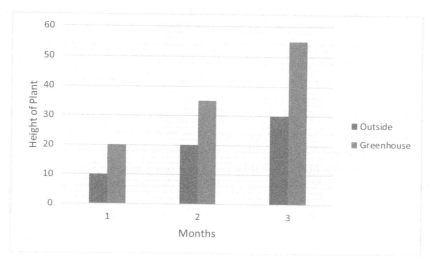

6. Looking at Figure 1, what gets trapped inside the greenhouse that helps plants grow?
    a. Short-wavelength IR
    b. Long-wavelength IR
    c. Cold air
    d. Water

7. Which plant grew taller from Scientist A's experiment?
    a. Outside
    b. Both grew to the same height.
    c. They both remained the same height for 3 months.
    d. Greenhouse

8. What gets converted to heat inside a greenhouse?
    a. Water
    b. Sunlight
    c. Plants
    d. Oxygen

9. What type of wavelength moves through the greenhouse glass easily according to Figure 1?
    a. Short-wavelength IR
    b. Oxygen
    c. Carbon dioxide
    d. Long-wavelength IR

10. What is one reason that plants may grow better inside a greenhouse?
    a. Colder air
    b. Less photosynthesis occurs in the greenhouse
    c. Fewer pests
    d. Less sunlight comes into the greenhouse

Passage 3

Questions 11–15 pertain to Passage 3:

In chemistry, a titration is a method that is used to determine the concentration of an unknown solution. Generally, a known volume of a solution of known concentration is mixed with the unknown solution. Once the reaction of the two solutions has been completed, the concentration of the unknown solution can be calculated. When acids and bases are titrated, the progress of the reaction is monitored by changes in the pH of the known solution. The equivalence point is when just enough of the unknown solution has been added to neutralize the known solution. A color reaction may also occur so that with the drop of solution that causes complete neutralization, the solution turns bright pink, for example. For acids that only have one proton, usually a hydrogen atom, the halfway point between the beginning of the curve and the equivalence point is where the amount of acid and base are equal in the solution. At this point, the pH is equal to the pK_a, or the acid dissociation constant.

Figure 1 below shows a general titration curve of a strong acid with a strong base.

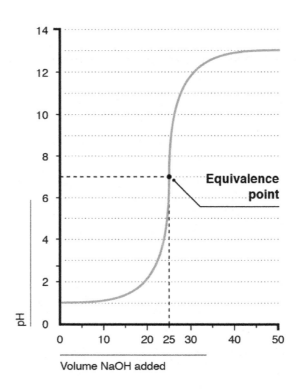

Figure 2 below shows the chemical reaction of a strong acid with a strong base.

Figure 3 shows the titration curve for acetic acid.

Titration Curve of Acetic Acid

OH⁻ Equivalents

11. How much NaOH is added to the HCl solution to reach the equivalence point in Figure 1?
    a. 10
    b. 40
    c. 50
    d. 25

12. What is the acid dissociation constant of the titration curve in the Figure 3?
    a. 4.21
    b. 3.50
    c. 4.76
    d. 6.52

13. What is the pH of the acetic acid before the titration has started in Figure 3?
    a. 2
    b. 4.76
    c. 7
    d. 6

14. What is one of the products of the chemical equation in Figure 2?
    a. HCl
    b. NaCl
    c. NaOH
    d. Cl⁻

15. How would you describe the solution at the equivalence point in Figure 1?
    a. Neutral
    b. Acidic
    c. Basic
    d. Unknown

Passage 4

Questions 16–20 pertain to Passage 4:

The heart is a muscle that is responsible for pumping blood through the body. It is divided into four chambers: the right atrium, right ventricle, left atrium, and left ventricle. Blood enters the atria and is then pumped into the ventricles below them. There is a valve between the atria and ventricles that prevents the blood from flowing back into the atria. The valve between the right atrium and ventricle has three folds whereas the valve between the left atrium and ventricle has two folds. Arteries carry oxygen-rich blood away from the heart to the body. Veins carry oxygen-poor blood from the body back to the heart. From there, the blood gets pumped to the lungs to get re-oxygenated and then back to the heart before circulating to the body. The heart beats every second of the day. For an adult, the normal heartrate is between 60 and 100 beats per minute. For a child, a normal heartrate is between 90 and 120 beats per minute.

Figure 1 below shows how blood gets pumped through the body.

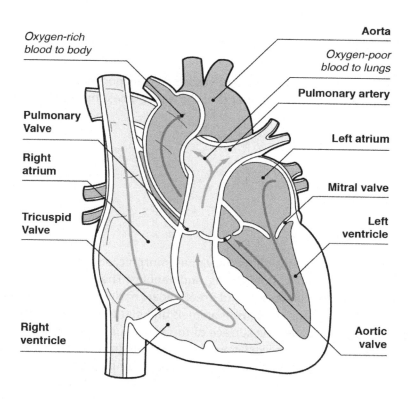

16. Where is the oxygen-poor blood pumped to before returning to the heart to get circulated to the rest of the body?
    a. Lungs
    b. Brain
    c. Stomach
    d. Kidney

17. Which heart valve has two folds?
    a. Pulmonary artery
    b. Tricuspid valve
    c. Mitral valve
    d. Aorta

18. If the aorta contains oxygen-rich blood, what type of vessel is it?
    a. Vein
    b. Pulmonary
    c. Airway
    d. Artery

19. Which heartrate (beats per minute) would be considered normal for a child during resting conditions?
   a. 85
   b. 125
   c. 100
   d. 60

20. The aorta is the artery that breaks into smaller vessels to transport blood to the rest of the body. Looking at the figure, which is the final chamber that the blood flows through before entering the aorta?
   a. Right ventricle
   b. Left ventricle
   c. Right atrium
   d. Left atrium

Passage 5

Questions 21–25 pertain to Passage 5:

There are three types of rocks: sedimentary, metamorphic, and igneous. Sedimentary rock is formed from sediment, such as sand, shells, and pebbles. The sediment gathers together and hardens over time. It is generally soft and breaks apart easily. This is the only type of rock that contains fossils, which are the remains of animals and plants that lived a long time ago. Metamorphic rock forms under the surface of the earth due to changes in heat and pressure. These rocks usually have ribbon-like layers and may contain shiny crystals. Igneous rock forms when molten rock, or magma, cools and hardens. An example of molten rock is lava, which escapes from an erupting volcano. This type of rock looks shiny and glasslike.

Figure 1 below is a chart of different types of rocks.

Igneous		Sedimentary		Metamorphic	
Obsidian	Pumice	Shale	Gypsum	Slate	Marble
Scoria	Rhyolite	Sandstone	Dolomite	Schist	Quartzite
Granite	Gabbro	Conglomerate	Limestone	Gneiss	Anthracite

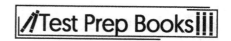

21. A volcano erupts and lava comes out and hardens once it is cooled. What type of rock is formed?
    A. Sedimentary
    B. Metamorphic
    C. Igneous
    D. Lava does not cool

22. Scientist A found a piece of granite rock, as seen in Figure 1. What type of rock is it?
    a. Igneous
    b. Metamorphic
    c. Sedimentary
    d. Fossil

23. Which type of rock could a fossil be found in?
    a. Igneous
    b. Bone
    c. Metamorphic
    d. Sedimentary

24. Which is an example of a metamorphic rock in the figure?
    a. Sandstone
    b. Slate
    c. Granite
    d. Limestone

25. What type of rock would most likely be formed at and found on the beach?
    a. Sedimentary
    b. Shells
    c. Igneous
    d. Metamorphic

Passage 6

Questions 26–30 pertain to Passage 6:

The greenhouse effect is a natural process that warms the Earth's surface, similar to what occurs in a greenhouse meant to grow plants. Solar energy reaches the Earth's atmosphere and warms the air and land. Some of the energy is absorbed by the greenhouse gases found in the Earth's atmosphere and by the land, and the rest is reflected back into space. Greenhouse gases include water vapor, carbon dioxide, methane, nitrous oxide, and chlorofluorocarbons. In recent decades, human activity has increased the amount of greenhouse gases present in the Earth's atmosphere, which has created a warmer atmosphere than normal and increased the Earth's temperature.

Figure 1 below shows the process of the greenhouse effect.

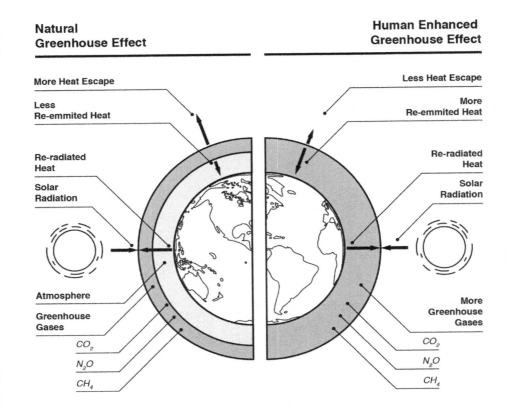

Figure 2 below describes the greenhouse gases that are produced from human activity.

Figure 3 below describes the human activities that produce carbon dioxide.

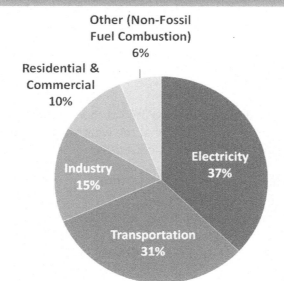

U.S. Carbon Dioxide Emissions, By Source

Other (Non-Fossil Fuel Combustion) 6%

Residential & Commercial 10%

Industry 15%

Electricity 37%

Transportation 31%

U.S. Environmental Protection Agency (2014).
U.S. Greenhouse Gas Inventory Report: 1990-2014.

26. Which is not an example of a greenhouse gas?
    a. Methane
    b. Carbon dioxide
    c. Nitrous oxide
    d. Helium gas

27. Looking at Figure 3, what could a person do to decrease how much carbon dioxide they produce?
    a. Leave lights on all the time
    b. Walk instead of drive a car
    c. Always drive in their own car everywhere
    d. Leave the television on all the time

28. Looking at Figure 2, which is the second highest greenhouse gas produced by human activity?
    a. Methane
    b. Fluorinated gases
    c. Nitrous oxide
    d. Carbon dioxide

29. Looking at Figure 1, what gets increasingly trapped in the Earth's atmosphere with increased human activity?
    a. Space
    b. The Sun
    c. Heat
    d. Water vapor

30. What type of charts are found in Figures 2 and 3?
    a. Scatter plots
    b. Line graphs
    c. Bar graphs
    d. Pie charts

Passage 7

Questions 31–35 pertain to Passage 7:

A meteorologist uses many different tools to predict the weather. They study the atmosphere and changes that are occurring to predict what the weather will be like in the future. Listed below are some of the tools that a meteorologist uses:

- Thermometer: measures air temperature
- Barometer: measures air pressure
- Rain gauge: measures rainfall over a specific time
- Anemometer: measures air speed
- Wind vane: shows which direction the wind is blowing

Figure 1 below shows data that is collected by a meteorologist.

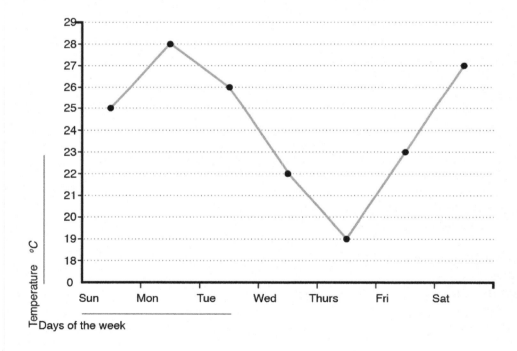

Figure 2 below shows data collected from a rain gauge.

Rainfall

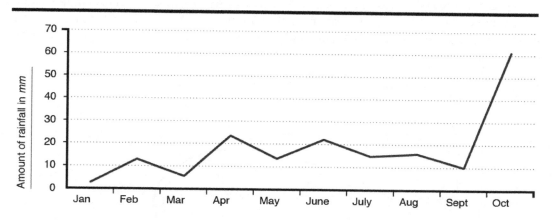

31. What tool would a meteorologist use to find out how fast the wind is blowing?
    a. Anemometer
    b. Barometer
    c. Thermometer
    d. Wind vane

32. What tool was used to collect the data shown in Figure 1?
    a. Rain gauge
    b. Thermometer
    c. Barometer
    d. Anemometer

33. The wind vane is pointing north. What does this tell us?
    a. Wind is blowing in an eastern direction.
    b. A storm is coming.
    c. The wind is blowing in a northern direction.
    d. The wind is blowing in a southern direction.

34. Looking at Figure 2, which month had the lowest rainfall?
    a. January
    b. April
    c. September
    d. October

35. Looking at Figure 2, what was the approximate amount of rain that fell in June?
    a. 0 mm
    b. 10 mm
    c. 50 mm
    d. 20 mm

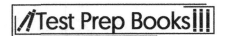

Passage 8

Questions 36–40 pertain to Passage 8:

Cells are the smallest functional unit of living organisms. Organisms can be single-celled or multicellular. Each cell contains organelles that are responsible for distinct functions and are essential for the organism's life. Plants and animals have different necessities for generating energy and nutrients. Their cells are similar but also have unique features.

Figure 1 below is a depiction of the organelles in an animal cell.

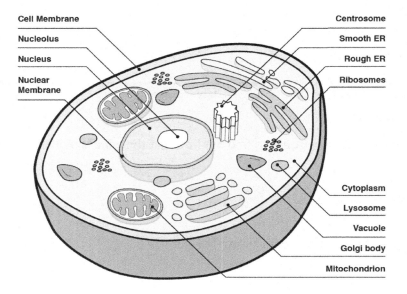

Figure 2 below depicts the organelles of a plant cell.

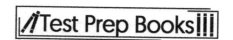

Figure 3 below describes the function of many important organelles.

Cell Organelle	Function
Cell wall	*(Plants only)* Maintains the shape of the cell and is a protective barrier for the internal contents of the cell.
Chloroplasts	*(Plants only)* Site of photosynthesis, which converts sunlight energy to glucose storage energy.
Nucleus	*(Plants and Animals)* Contains the cell's DNA.
Ribosomes	*(Plants and Animals)* Puts together long chains of amino acids to build proteins. Smallest organelle in the cell.
Mitochondria	*(Plants and Animals)* The powerhouse of the cell. Converts the stored glucose energy to ATP energy, which drives forward almost all of the cell's reactions.
Cell membrane	*(Plants and Animals)* Regulates what molecules can move in and out of the cell. Made of a phospholipid bilayer.
Cytoplasm	*(Plants and Animals)* The liquid that fills the inside of the cell.
Vacuole	*(Plants and Animals)* A membranous sac that encloses anything in the cell that needs to be kept separate, such as food and water.
Golgi Body	*(Plants and Animals)* Receives products produced by the endoplasmic reticulum (ER) and adds final changes to them.
Lysosomes	*(Plants and Animals)* A membranous sac that is full of digestive juices. Breaks down larger molecules into smaller parts so that they can be used to build new parts of the cell.
Rough endoplasmic reticulum (ER)	*(Plants and Animals)* A large folded membrane that is covered with ribosomes. Helps fold and modify the proteins built by the ribosomes before sending them to the Golgi body.
Smooth endoplasmic reticulum (ER)	*(Plants and Animals)* A large folded membrane that puts together lipids.
Microtubules and microfilaments	*(Plants and Animals)* Long tubes that allow the cell to move and provide an internal structure of support for the cell.

36. Which is an organelle found in a plant cell but not an animal cell?
    a. Mitochondria
    b. Chloroplast
    c. Golgi body
    d. Nucleus

37. Where is the nucleolus located in both plant and animal cells?
    a. Near the chloroplast
    b. Inside the mitochondria
    c. Inside the nucleus
    d. Attached to the cell membrane

38. Which organelle is responsible for generating energy for the cell and is referred to as the powerhouse of the cell?
    a. Mitochondria
    b. Nucleus
    c. Ribosomes
    d. Cell wall

39. What does the cell membrane do?
    a. Builds proteins
    b. Breaks down large molecules
    c. Contains the cell's DNA
    d. Controls which molecules are allowed in and out of the cell

40. What are chloroplasts responsible for in plant cells?
    a. Maintaining the cell's shape
    b. Containing the cell's DNA
    c. Converting energy from sunlight to glucose
    d. Building proteins

Writing

Please read the prompt below and answer in an essay format.

Coaches of kids' sports teams are increasingly concerned about the behavior of parents at games. Parents are screaming and cursing at coaches, officials, players, and other parents. Physical fights have even broken out at games. Parents need to be reminded that coaches are volunteers, giving up their time and energy to help kids develop in their chosen sport. The goal of kids' sports teams is to learn and develop skills, but it's also to have fun. When parents are out of control at games and practices, it takes the fun out of the sport.

1. Analyze and evaluate the passage given.

2. State and develop your own perspective.

3. Explain the relationship between your perspective and the one given.

Answer Explanations for Practice Test #2

English

1. C: Choice C is correct, changing *Mythologies* to *mythology's*. Since one myth system is being referred to—and one particular component of it—the possessive is needed. Additionally, *Mythology's* does not need to be capitalized, since only the culture represents a proper noun. Choice A therefore is incorrect, with Choice B failing to fix the plural and Choice D having extraneous capitalization.

2. A: Choice A is correct because the sentence has no issues. While Choice B separates the sentence correctly, it makes more sense in this context of a direct comparison to keep the sentence intact. Choice C is incorrect because the sentence needs a comma after *superhero*. Choice D is unnecessarily long and lacks the word *but* that helps the author differentiate ideas.

3. B: Choice B is correct because it adds the two commas needed to clarify key subjects individually and establish a better flow to the sentence. Since *destroying mountains*, *defeating scores of giants*, and *lifting up the world's largest creature* are separate feats, commas are needed to separate them. Also, because *the world's largest creature* can stand alone in the sentence, a comma needs to proceed its name; *the Midgard Serpent* is not necessary to the sentence but rather provides extra information as an aside. Choice A is unclear and thus incorrect. Choice C is still missing a comma, while Choice D put an extraneous one in an incorrect place.

4. D: Choice D is correct since the sentence is lengthy as originally presented and should be split into two. Additionally, *however*, being a conjunction, needs a comma afterwards. Choice A is therefore incorrect due to missing punctuation. Choice B is an improvement but could separate the sentence's ideas better and more clearly. Choice C lacks the necessary comma after *However*.

5. A: Choice A is correct because this sentence has no issues with punctuation, content, or sentence construction. While there are three commas used, they serve to appropriately introduce an idea, an individual person, and transition into another line of thinking. Choices B and C miss commas needed to offset Loki's title as *the god of mischief*, while Choice D misses the comma needed to introduce the example.

6. C: Choice C is correct because the sentence needs two commas to emphasize the proper name of Mjolnir. Since Mjolnir is being talked about, directly addressed, and then explained, it must be flanked by commas to signify its role in the sentence. Choice A lacks necessary punctuation and is confusing. Choices B and D miss commas on either side of *Mjolnir*.

7. A: Choice A is correct, as this is an example of a compound sentence written correctly. Because of the conjunction *but* and the proceeding comma, the two independent clauses are able to form a single sentence coherently. While Choice B makes the question more direct, it doesn't go well with the remainder of the sentence. Choice C applies a comma after *but*, which is incorrect and confusing. Choice D inserts *however*, which is out of place and makes the sentence awkward.

8. B: Choice B correctly changes *an* to *a*, since *an* is only required when *a* precedes a word that begins with a vowel. Choice A therefore uses the incorrect form of *a*. Choice C fixes the issue but unnecessarily reverses the structure of the sentence, making it less direct and more confusing. Choice D does not fix the error and adds extraneous commas.

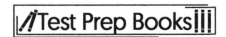

9. D: Choice *D* is correct, simply applying a comma after *Well* to introduce an idea. Choice *A* is therefore incorrect. Choice *B* introduces too many commas, resulting in a fractured sentence structure. Choice *C* applies a comma after *Well*, which is correct, but interrupts the flow of the sentence by switching the structure of the sentence. This makes the sentence lack fluidity and serves to confuse the reader.

10. C: Choice *C* is simple and straightforward, describing the event clearly for the reader to follow; *talked* is past tense, which is consistent with the rest of the passage. Choice *A* is incorrect, since *we came to talking about* confuses the tense of the sentence and the verb *talk*. Choices *B* and *D* are wordy and not as straightforward as Choice *C*.

11. B: Choice *B* is the correct answer because it adds a comma after *Essentially* and changes *an* to *a*. This is called the indefinite article, when an unspecified thing or quantity is referred to. However, *an* doesn't agree with *metaphor*, since *an* should only be used when the next word starts with a vowel. Choice *A* uses the article *an* and lacks the crucial comma after *Essentially*. Choice *C* is incorrect because it only provides the comma after *Essentially*, neglecting the indefinite article disagreement. Choice *D* is incorrect because neither issue is fixed and an unnecessary comma is introduced.

12. A: Choice *A* is correct because there are no errors present in the sentence. Choice *B* is a run on, because the clauses are not broken up by commas. Choice *C* has a verbal disagreement: *look* and *saw* are different tenses. Choice *D* changes the structure of the sentence but fails to add a transition to make this correct.

13. D: Choice *D* is correct because it uses a comma after the word *passage*, successfully connecting the dependent clause with its independent clause to form a complete thought/sentence. Choice *A* is therefore incorrect. Choice *B* uses a semicolon unwisely. The two clauses need to be connected to each other in order to make sense, otherwise they are just two fragments improperly combined. Choice *C* does not have the required comma and changes *found* to *finding*, an inappropriate tense for the verb in this sentence.

14. C: Choice *C* is correct because it fixes two major flaws in the original portion of the sentence. First, it inserts the adverb *also* to show the connection between *whatever it is that makes us unique* and *defines us*. Without this adverb, the sentence lacks clarity, and the connection is lost. Second, *big* is incorrect in this context. The sentence needs the superlative *bigger* in order to communicate the scope and scale of the author's assessment of how people relate to others on a grand scale. Choice *A* is therefore incorrect, Choice *B* inserts unnecessary commas, and Choice *D* subtly alters the original meaning.

15. B: Choice *B* is correct because a comma is correctly inserted after *ultimately*. This serves to express a side thought that helps transition into the rest of the sentence without having to break it apart. Choice *A* is incorrect because it lacks the comma after *ultimately*. Choice *C* uses too many commas and is overly complicated. Choice *D* lacks the necessary conjunction after the comma (*but*) before *ultimately*, making it a run-on sentence. It also lacks the important comma after *ultimately*.

16. D: Choice *D* corrects the subject-verb disagreement. *One's* is the possessive form of *one*, a single individual, not the plural *individuals*. *Isn't* is the singular contraction of *is not*, which conflicts with *individuals*. To correct this, either *isn't* must change to *aren't* or *individuals* should become the singular *individual*. The latter is correct because of the context of the sentence. Choice *A* is incorrect because of the subject-verb disagreement. Choice *B* uses the future tense, while Choice *C*'s *aren't* conflicts with *one's*, which is possessive singular.

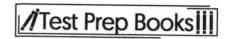

17. A: Choice *A* contains no grammatical errors and communicates the writer's message clearly. Choice *B* inserts an unnecessary semicolon. Choice C uses *is*, which disagrees with the plural *aspects*. *Are* must be used because it is plural. This is the same for Choice *D*, which uses *there's* (*there is*).

18. D: Choice *D* is the correct answer because it removes the comma after *Rushdie*. Adding a comma after the proper name in this case is incorrect because *Rushdie* is not being addressed directly. Rather, the writer is talking about Rushdie. Therefore, Choices *A* disrupts the construction of the sentence. Choice *B* is incorrect because *seemed*, in this context, should be present tense. The author is talking about a theme and idea that *Rushdie* had but that is still relevant and being actively studied. Choice *C* fails to remove the comma after Rushdie and applies the gerund *seeming* incorrectly.

19. A: Choice *A* is the correct answer. The sentence requires no punctuation and clearly communicates the author's idea. Choice *B* is incorrect because it misuses multiple commas. Choice *C* is incorrect because it uses *a* instead of *an*, which is necessary because the next word (*essentially*) begins with a vowel. Choice *D* changes *rewarding* to *rewarded*, which clashes with the earlier use of *was*, indicating past tense, making it incorrect.

20. B: Choice *B* is correct because it uses the present tense of *understand* instead of the past tense *understood*. *Continues* emphasizes something ongoing. Therefore, the present tense of *understand* is needed. Choice *A* therefore has tense disagreement. Choice *C* uses an extraneous comma. While Choice *D's* use of the gerund is a better option, *in* would need to be added before *understanding* for correctness.

21. A: Choice *A* is correct; the dependent and independent clauses are successfully combined to form a sentence. Choice *A* is also the most concise and straightforward option, presenting the information appropriately so as not to be confusing. Choice *B* can be eliminated because it changes the meaning of the sentence. Choice *C* unnecessarily flanks *exactly* with commas. Choice *D* is incorrect because a period after the underlined phrase would result in an incomplete sentence.

22. D: Choice *D* is correct because of how it successfully connects the two sentences. By starting a new sentence with *instead*, the two ideas are clearly and correctly presented. Choice *A* is a run-on. Choice *B* is incorrect because the commas used are misplaced and confusing. Choice *C* is incorrect because there needs to be a comma after *originally*.

23. B: Choice *B* correctly uses the word *passed* instead of *passing*. The word *were* is being used as a past tense modifier; therefore, it disagrees with the *passing* participle. Combining the helping verb *were* with the past tense *passed* creates the correct past tense compound verb, *were passed*, which is needed to be grammatically consistent with the rest of the sentence. Choices *A*, *C*, and *D* thus have incongruent tenses.

24. D: Choice *D* is correct because it corrects the subject-verb disagreement between the plural *myths* and the singular *it*. This is done by changing *it* to *them*, reflecting that more than one type of *myths* are being talked about. Because of the subject-verb disagreement, Choice *A* is incorrect. Choice *B* does not correct this at all. Choice *C* addresses this disagreement, but at the cost of maintaining parallel structure within the sentence.

25. A: Choice *A* is correct because this sentence contains no grammatical errors, unlike the others. Choice *B* uses the semicolon when only a comma is required. Choice *C* uses commas to unnecessarily isolate *a kind of soundtrack* from its connected phrase. Choice *D* is incorrect because the plural *kinds* disagrees with the singular pronoun *This*.

26. C: Choice *C* is the correct answer because the plural *are* and *motion pictures* agree with the subject *stimuli* (the plural form of *stimulus*). All verbs and nouns in the sentence should be in agreement with each other in tense and number. Choices *A*, *B*, and *D* all violate this number agreement.

27. B: Choice *B* correctly replaces the preposition *from* with the correct proposition *to*. *To* is used to emphasize an action or idea being presented rather than received. This is why Choice *A* is incorrect. Choice *C* is incorrect because it uses the adverb *too*, which emphasizes a higher level of something or the addition of something in a sentence; this isn't relevant here. Choice *D* uses the preposition *for* instead of *to*.

28. D: Choice *D* is correct because it adds the helping verb *were* to modify *unsettled*. This allows the sentence to reflect that the founders *were unsettled* by the implications. Without *were* to connect the founders to *unsettled*, the sentence doesn't make sense. Choice *A* lacks the crucial helping verb, making it incorrect. Choice *B* is incorrect because of its unnecessary semicolon. Choice *C* changes *theory's*, which is singular possessive, to *theories'* (plural possessive), which isn't consistent with the sentence's context.

29. A: Choice *A* is correct because it contains no errors and requires no additional punctuation to form a coherent sentence. The single comma, used successfully, unites the two clauses and enables a solid grammatical structure. Choice *B* incorrectly places a comma after *Instead*, Choice *C* incorrectly changes *predicted* to *predicting*, and Choice *D* incorrectly separates *where everything can be predicted by equations* from the rest of the sentence.

30. D: Choice *D* is correct because it fixes the subject-verb disagreement with *Every outcome* and *exist*. *Exists* is third person present but also appropriate to reflect multiple outcomes, as indicated by *every outcome*. Choices *A* and *B* use *exist*, not *exists*, which makes them both incorrect. Choice *C* is fine on its own but does not fit with the rest of the sentence.

31. B: Choice *B* is correct because the comma after *determinism* isn't needed. Adding a comma in the selected area actually breaks up the independent clause of the sentence, thus compromising the overall structure of the sentence. Choices *A*, *C*, and *D* are therefore incorrect.

32. C: Choice *C* is the correct answer because it removes the contraction of *it is*, *it's*. Choice *A*, which is incorrect, originally used *it's*—note the apostrophe before *s*. *It's* simply means *it is*, while *its* (no apostrophe) shows possession. In this sentence, *its* is referring to the idea devised by Schrödinger, giving ownership of the purpose to the idea. Choice *B* is incorrect because *purpose* should remain singular. Choice *D* is incorrect because it uses *it's*.

33. A: Choice *A* is correct because the sentence is well-formed and grammatically correct. Choice *B* is incorrect because it adds an unnecessary comma after *if*. Choice *C* breaks the sentence apart, creating a sentence fragment. Choice *D* is incorrect because it changes *release* to a gerund and fails to make a coherent sentence, leaving only two dependent clauses.

34. D: Choice *D* is the correct answer. This is a tricky question, but Choice *D* is correct because, in the context of this sentence, it's important to have *however* flanked by commas. This is because the use of *however* is basically an aside to the reader, addressing an idea and then redirecting the reader to an alternative outcome or line of reasoning. Choice *A* is therefore confusing, with *however* floating in the sentence aimlessly. Choice *B* only uses one comma, which is incorrect. Choice *C* creates two incomplete sentences.

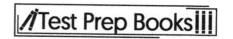

35. A: Choice *A* is correct. The sentence uses a lot of commas, but these are used effectively to highlight key points while continuing to focus on a central idea. Choice *B* is incorrect because the commas after *reality* and *that* are required. Choice *C* is incorrect because there should be a comma after *that* because *on the smallest scales* elaborates on the idea itself but not necessarily what Broglie said. Choice *D* puts a comma after *further*, which is unnecessary in this context.

36. B: Choice *B* correctly uses *an* instead of *a* to modify *interference*. The indefinite article *an* must be used before words that start with a vowel sound. The verb *created* is also in agreement with the tense of the story. Choice *C* incorrectly changes *creates* to *create* and pluralizes *light*, which is inconsistent with the rest of the sentence. Choice *D* modifies *creates* inappropriately and adds an incorrect comma after *light*.

37. C: Choice *C* is correct because it applies the single comma needed to combine the independent clause with the dependent clause, forming a functional sentence. It also includes quotes. Choice *A* is an awkward, disjointed sentence because it lacks a comma after *goes*. The other answers result in oddly constructed sentences (Choice *B*), and a misused comma after *As* (Choice *D*).

38. A: Choice *A* is the correct answer because the proper contraction for does not (doesn't) is used. There is also no need for commas or punctuation here as well, making Choices *B* and *C* incorrect. Choice *D* needlessly changes *doesn't* to *aren't*.

39. B: Choice *B* is correct because it fixes the subject-verb disagreement between the subject *the basic knowledge* and the verb *are*. Because *knowledge* is singular, the verb *are*, which is plural, is incorrect. Instead, the singular *is* must be used in place of *are*. This makes Choices *A* and *C* incorrect. Choice *D* is incorrect because it changes the meaning of the sentence.

40. C: Choice *C* is the correct answer because it employs a comma to effectively combine the independent clause with the dependent clause to form a complete sentence. Choice *A* lacks the comma after *thinking* needed to unite the two parts of the sentence. The independent clause is *Smart leaders will engage in critical thinking*, while the independent clause is *allowing them to discern ulterior motives and identify propaganda*. Choice *B* incorrectly adds a comma after *motives*. Choice *D* is incorrect because it still lacks connective punctuation and incorrectly alters the gerund *allowing* to *allows*; this would be fine if a comma and *which* were added before *allows* to effectively combine the clauses.

41. D: Choice *D* is the best answer because it changes *these* to *this*, making it properly modify the singular *practice*. Choice *A* is incorrect because of this lack of numerical agreement. Choice *B* is incorrect for the same reason—*this* is singular while *practices* is plural. Choice *C* is also incorrect because it doesn't fix *these* and takes out *will*, which is important for the tense of the overall sentence.

42. A: Choice *A* is correct because the sentence is fine as it is. Commas are not necessary here. Choices *B* and *C* replace *to* with *for* and *from*, respectively, which don't work in the context of the sentence: if leaders are *receptive*, they are receiving something, so *to* is appropriate. Choice *D* introduces unnecessary commas.

43. B: Choice *B* is the correct answer because it appropriately applies a comma after the opening word *Sadly*. This is because the author is introducing an idea or feeling, then transitioning into an elaborative explanation. *Sadly* is an aside, so there must be a comma afterwards to transition between thoughts and avoid a long-winded sentence. Choice *A* lacks the comma, so it is incorrect. Choice *C* incorrectly separates *that are attacked as lies* from the rest of the sentence, breaking its flow. Choice *D* still lacks the comma and incorrectly changes the two past tense verbs, *plagued* and *attacked*, into gerunds.

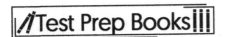

44. D: Choice *D* is the correct answer because the tense of *understanding* best applies to the context of the sentence. *Understanding* is also preceded by the modifier *an*. The combined *an understanding* forms a compound noun that is the direct object and focus of the sentence. Choices *A* and *B* use *understand* as a verb, which doesn't complement the sentence. Choice *C* lacks the modifying *an*.

45. A: Choice *A* is the best option. There are no errors in the original sentence, as both parts of the sentence (dependent and independent clauses) are combined through the central comma. Choice *B* creates two new sentences, the first of which is not complete with *While* still present. Choice *C* introduces an unnecessary comma after *practices*. Choice *D* uses *whom* instead of *who*, which is the correct form for this sentence.

46. B: Choice *B* is the correct answer because it corrects the subject-verb disagreement between *character* and *retain* by using *retains* instead. *Retain* is plural, which clashes with the singular *character*, making Choice *A* incorrect. *Retains* is the correct form because it is third person singular, meaning that Choice *C* is also incorrect. Choice *D* is possible as a revision but does not agree with the tense of *do take on* earlier in the sentence.

47. D: Choice *D* is the correct answer because it uses a comma to link the two halves of the sentence and adds the phrase *as well as* to help transition into the dependent clause. *As well as* is also used to communicate that *The Canterbury Tales* shares other qualities besides being *a story about stories*. Choice *A* is incorrect because it is a run on. In particular, *stories also being a story* is very unclear, necessitating the swap of *also* for *as well as*. Choice *B* has an awkward construction and uses the semicolon unwisely; this is certainly not the best option. Choice *C* uses the correct comma but lacks the phrase *as well as*.

48. C: Choice *C* is the correct answer because *these* is the incorrect pronoun for this particular section of the sentence; instead, the singular pronoun *this* should be used. Therefore, Choice *A* is incorrect. While Choice *B* correctly uses *this*, it also inserts a comma afterwards, which disrupts the flow of the sentence. Choice *D* is incorrect because it uses *these* and incorrectly uses *through* instead of *by*.

49. A: Choice *A* is correct. The original sentence has no issues and the author's thoughts are communicated clearly. Choice *B* incorrectly separates the sentence into two sentences, which creates a sentence fragment. Choice *C* rearranges the section in a way that clashes with the rest of the sentence. Choice *D* is incorrect because of the comma inserted after *clear*, breaking up the sentence unnecessarily.

50. B: Choice *B* is the correct answer because it corrects the subject-verb disagreement within the underlined portion. *These* is plural, so it must modify a plural noun. Thus, the plural *individuals* is required. Another clue for this is the fact that *descriptions* is also plural. Choice *A* is therefore incorrect. Choice *C* removes the necessary comma, while Choice *D* inserts an unnecessary semicolon.

51. D: Choice *D* is correct for two reasons. The incorrect pronoun *her* is replaced with the appropriate pronoun *she*. More specifically, Choice *D* uses *she's*, the contraction of *she is*. This emphasizes that the subject, *she*, is acting in a certain way or doing something. Clearly, the author is describing the nun's actions or characteristics. Choice *A* is incorrect because *her* is an inappropriate pronoun in this context, and there is no form of *is* to express what the nun is doing. Choices *B* and *C* still use *her*, so they are incorrect, with the former also altering the sentence completely.

52. C: Choice *C* is the answer, correctly changing the adjective *brutal* to the adverb *brutally* so it can properly modify the adjective *honest*. Without this change, the sentence makes no sense because it presents a stream of adjectives that don't cleanly fit together. Thus, Choice *A* is incorrect. Choice *B* introduces too many commas and remains incoherent, while Choice *D* fails to modify *brutal*.

53. B: Choice *B* is the correct choice because it fixes the confusion of *shed's*. In this context, *shed* is not used as a noun but as a verb, and a verb cannot be possessive. What the sentence should have is *sheds*, the third-person simple singular present form of the verb to complete the phrase: *sheds light on*. This means Choice *A* is incorrect. Choice *C* alters the section in a way that clashes with the whole sentence. Choice *D* fails to correct *shed's* and introduces an unnecessary comma.

54. A: Choice *A* is the correct answer because there are no issues with this section of the sentence. The whole sentence is clear and uses proper punctuation. Choice *B* lacks an introductory comma after *likewise*, which is incorrect. Choices *C* and *D* add unnecessary commas.

55. B: Choice *B* is correct because it makes *Practitioners* plural, enabling it to agree with the verb *strive*. Making *Practitioners* plural also matches the context of the sentence. Therefore, Choice *A* is incorrect. Choice *C* is incorrect because it takes away a necessary comma. Choice *D* makes *Practitioner* and *strives* both singular, but without a modifier like *A* or *The* in front of *practitioner*, the sentence is not correct.

56. D: Choice *D* is the best answer because it corrects the two key errors in the section. *Them* is not the proper pronoun to use in this sentence. *They*, the third-person plural personal pronoun, should be used instead. Choices *A* and *B* are thus incorrect. Choice *D* also adds *be* to modify the main verb *used*, forming the appropriate compound verb. Choice *C* is missing this essential component.

57. C: Choice *C* is correct because the sentence requires the presence of *a* to introduce and modify the noun *testament*. *Testament* is a noun meaning evidence of quality or belief; it's not a verb, which is how Choice *A* uses it. This makes Choice *A* incorrect. Choice *B* uses *an* instead of *a*, which is not necessary because *testament* begins with a consonant. Choice *D* adds an unneeded comma after *possibilities*.

58. D: Choice *D* is the answer because it applies the two commas needed around *like any trial throughout one's life* in order to introduce a side thought in the sentence clearly, without breaking up the flow. The direct object is *test*. Therefore, the commas implemented before and after the phrase are required. Choice *A* doesn't contain the comma after *life*. Choice *B* lacks both the needed commas. Choice *C* applies the second comma incorrectly after *like*, which disrupts the construction of the sentence.

59. A: Choice *A* is correct because it contains no errors that mar the grammar or flow of the sentence. Choice *B* is incorrect because *upon* is incorrectly used to replace *on*. *Upon* refers specifically to a surface, which is not appropriate for this sentence. The preposition *on* is needed here. Choice *C* alters the sentence unnecessarily, confusing the tense and focus of the sentence by using *looking* instead of *look*. Choice *D* changes *some* to *something*, which makes no sense for the rest of the sentence.

60. C: Choice *C* is the correct answer because it adds a comma after *Certainly*. This is important because the author is addressing the audience before moving on to explore *his ideas*. Choice *A* is incorrect because the lack of a comma makes this sentence a run-on. Choices *B* and *D* are incorrect because, while the former applies the comma after *Certainly*, they both take away the *but* that modifies *at what costs*.

61. B: Choice *B* corrects the verb tense of *focuses*. The phrase *to focuses* is not an appropriate infinitive; the best replacement is *to focus*. The sentence is not providing the reason why *Plato's ideal society is one that places human desire*, so Choice *A* is incorrect. Using *focus* allows the prepositional phrase *to*

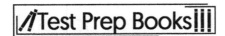

focus on to fluidly transition into the second half of the sentence. Choice *C* is not a bad option, but it lacks the comma needed to transition into *focusing on*. Choice *D* incorrectly eliminates *to*.

62. C: Choice *C* is the correct answer because it replaces the possessive *its* with the contraction *it's* and adds a transition to an alternate viewpoint. *But* should be added as a conjunction after the comma, linking the dependent clause and independent clause. Choices *A* and *B* are therefore incorrect. Choice *D* adds a transition but incorrectly removes *its*, rendering the sentence incoherent.

63. D: Choice *D* is the correct answer because it adds an apostrophe to *citizens*, making it *citizens'*. This gives ownership to *citizens*, which clearly indicates the idea of many citizens having their children confiscated. Choices *A*, *B*, and *C* are all incorrect because they don't give ownership to citizens: the latter adds the apostrophe in a place that indicates only one citizen is impacted.

64. B: Choice *B* is correct because it changes the plural *these* to the singular *this* in order to agree with the singular noun *kidnapping*. Choice *A* is incorrect because of this number disagreement. Choice *C* would be a good alternative, but it lacks the adverb *as* to clarify the use of *kidnappings* and eliminates *and*, which serves as the connection to the rest of the sentence. Choice *D* is incorrect because *those* is plural while *kidnapping* remains singular.

65. A: Choice *A* is correct because the sentence lacks grammatical errors and successfully joins the dependent clause with the independent clause by inserting a single comma after *beings*. Choice *B* is incorrect because it needs a comma to connect the dependent clause and independent clause. Choice *C* uses a semicolon instead of a simple comma. Choice *D* is incorrect because the plural *those* replaces *this*. *This* is singular to describe a single event, which agrees with *seems*.

66. D: Choice *D* is the correct choice because it fixes two issues with the underlined section that cause confusion for the whole sentence. First, *than* is incorrectly used in the original section. *Than* is a conjunction used to make a comparison, while *then* (acting as an adverb) serves to express a result of something, like a sequence of events. Therefore, Choices *A* and *B* are incorrect. However, therein lies the trickiness of the question because none of the answer choices use *then* without altering the meaning of the sentence (Choice *C*). Choice *D* compensates for this by replacing *than/then* with a simple comma to link the two clauses and transition from the central idea to the potential outcome of the idea (the child having more opportunities).

67. C: Choice *C* corrects the only flaw in this sentence by fixing the tense of *encourage*. *Encourages* is the third person present form of the original verb and as such fits much better because the sentence is written in the third person. Choice *A* is incorrect because the tense of *encourage* doesn't sync with the rest of the sentence. Choice *B* is incorrect; the gerund form of *encourage* is unnecessary. Choice *D* uses the conjunction *which* instead *that*, which is incorrect, in addition to not fixing the tense of *encourage*.

68. B: Choice *B* is correct. Here, a colon is used to introduce an explanation. Colons either introduce explanations or lists. Additionally, the quote ends with the punctuation inside the quotes, unlike Choice *C*.

69. A: The verb tense in this passage is predominantly in the present tense, so Choice *A* is the correct answer. Choice *B* is incorrect because the subject and verb do not agree. It should be *Education provides*, not *Education provide*. Choice *C* is incorrect because the passage is in present tense, and *Education will provide* is future tense. Choice *D* doesn't make sense when placed in the sentence.

70. D: The possessive form of the word *it* is *its*. The contraction *it's* denotes *it is*. Thus, Choice *A* is incorrect. The word *raises* in Choice *B* makes the sentence grammatically incorrect. Choice *C* adds an apostrophe at the end of *its*. While adding an apostrophe to most words would indicate possession, adding *'s* to the word *it* indicates a contraction.

71. C: The word *civilised* should be spelled *civilized*. The words *distinguishes* and *creatures* are both spelled correctly.

72. B: Choice *B* is correct because it provides clarity by describing what *myopic* means right after the word itself. Choice *A* is incorrect because the explanation of *myopic* comes before the word; thus, the meaning is skewed. It's possible that Choice *C* makes sense within context. However, it's not the best way to say this because the commas create too many unnecessary phrases. Choice *D* is confusingly worded. Using *myopic focus* is not detrimental to society; however, the way *D* is worded makes it seem that way.

73. C: Again, we see where the second paragraph can be divided into two parts due to separate topics. The paragraph's first main focus is education addressing the mind, body, and soul. This first section, then, could end with the concluding sentence, "The human heart and psyche . . ." The next sentence to start a new paragraph would be "Education is a basic human right." The rest of this paragraph talks about what education is and some of its characteristics.

74. A: Choice *A* is correct because the phrase *others' ideas* is both plural and indicates possession. Choice *B* is incorrect because *other's* indicates only one *other* that's in possession of *ideas*, which is incorrect. Choice *C* is incorrect because no possession is indicated. Choice *D* is incorrect because the word *other* does not end in *s*. Others's is not a correct form of the plural possessive word.

75. D: This sentence must have a comma before *although* because the word *although* is connecting two independent clauses. Thus, Choices *B* and *C* are incorrect. Choice *A* is incorrect because the second sentence in the underlined section is a fragment.

Math

1. A: Compare each numeral after the decimal point to figure out which overall number is greatest. In answers *A* (1.43785) and *C* (1.43592), both have the same tenths (4) and hundredths (3). However, the thousandths is greater in answer *A* (7), so *A* has the greatest value overall.

2. D: By grouping the four numbers in the answer into factors of the two numbers of the question (6 and 12), it can be determined that:

$$(3 \times 2) \times (4 \times 3) = 6 \times 12$$

Alternatively, each of the answer choices could be prime factored or multiplied out and compared to the original value.

6×12 has a value of 72 and a prime factorization of $2^3 \times 3^2$.

The answer choices respectively have values of 64, 84, 108, 72, and 360, so Choice *D* is correct.

3. C: The sum total percentage of a pie chart must equal 100%. Since the CD sales take up less than half of the chart and more than a quarter (25%), it can be determined to be 40% overall. This can also be measured with a protractor. The angle of a circle is 360°. Since 25% of 360° would be 90° and 50% would be 180°, the angle percentage of CD sales falls in between; therefore, it would be Choice *C*.

4. B: Using the conversion rate, multiply the projected weight loss of 25 lb. by 0.45 $\frac{kg}{lb}$ to get the amount in kilograms (11.25 kg).

5. D: First, subtract $1437 from $2334.50 to find Johnny's monthly savings; this equals $897.50. Then, multiply this amount by 3 to find out how much he will have (in three months) before he pays for his vacation: this equals $2692.50. Finally, subtract the cost of the vacation ($1750) from this amount to find how much Johnny will have left: $942.50.

6. C: Follow the *order of operations* in order to solve this problem. Solve the parentheses first, and then follow the remainder as usual:

$$(6 \times 4) - 9$$

This equals 24 − 9 or 15, answer *C*.

7. D: Three girls for every two boys can be expressed as a ratio: 3:2. This can be visualized as splitting the school into 5 groups: 3 girl groups and 2 boy groups. The number of students which are in each group can be found by dividing the total number of students by 5:

$$\frac{650 \text{ students}}{5 \text{ groups}} = \frac{130 \text{ students}}{\text{group}}$$

To find the total number of girls, multiply the number of students per group (130) by the number of girl groups in the school (3). This equals 390, Choice *D*.

8. C: Kimberley worked 4.5 hours at the rate of $10/h and 1 hour at the rate of $12/h. The problem states that her pay is rounded to the nearest hour, so the 4.5 hours would round up to 5 hours at the rate of $10/h:

$$(5h)(\$10/h) + (1h)(\$12/h) = \$50 + \$12 = \$62$$

9. C: The first step is to depict each number using decimals. $\frac{91}{100}$ = 0.91

Dividing the numerator by denominator of $\frac{4}{5}$ to convert it to a decimal yields 0.80, while $\frac{2}{3}$ becomes 0.66 recurring. Rearrange each expression in ascending order, as found in Choice *C*.

10. B: First, calculate the difference between the larger value and the smaller value:

$$378 - 252 = 126$$

To calculate this difference as a percentage of the original value, and thus calculate the percentage *increase*, divide 126 by 252, then multiply by 100 to reach the percentage 50%, Choice *B*.

11. E: Simplify each mixed number of the problem into a fraction by multiplying the denominator by the whole number and adding the numerator:

$$\frac{14}{3} - \frac{31}{9}$$

Since the first denominator is a multiple of the second, simplify it further by multiplying both the numerator and denominator of the first expression by 3 so that the denominators of the fractions are equal:

$$\frac{42}{9} - \frac{31}{9} = \frac{11}{9}$$

Simplifying this further, divide the numerator 11 by the denominator 9; this leaves 1 with a remainder of 2. To write this as a mixed number, place the remainder over the denominator, resulting in $1\frac{2}{9}$.

12. C: We are trying to find x, the number of red cans. The equation can be set up like this:

$$x + 2(10 - x) = 16$$

The left x is actually multiplied by \$1, the price per red can. Since we know Jessica bought 10 total cans, $10 - x$ is the number blue cans that she bought. We multiply the number of blue cans by \$2, the price per blue can.

That should all equal \$16, the total amount of money that Jessica spent. Working that out gives us:

$$x + 20 - 2x = 16$$

$$20 - x = 16$$

$$x = 4$$

13. C: Janice will be choosing 4 employees out of a set of 6 applicants, so this will be given by the choice function. The following equation shows the choice function worked out:

$$\binom{6}{4} = \frac{6!}{4!\,(6-4)!} = \frac{6!}{4!\,(2)!}$$

$$\frac{6 \times 5 \times 4 \times 3 \times 2 \times 1}{4 \times 3 \times 2 \times 1 \times 2 \times 1} = \frac{6 \times 5}{2} = 15$$

14. D: $\frac{3}{100}$. Each digit to the left of the decimal point represents a higher multiple of 10 and each digit to the right of the decimal point represents a quotient of a higher multiple of 10 for the divisor.

The first digit to the right of the decimal point is equal to the value ÷ 10. The second digit to the right of the decimal point is equal to the value ÷ (10×10), or the value ÷ 100.

15. D: Using the order of operations, items inside of parentheses are sorted out first. Therefore, the teacher should resolve the parentheses first. In this expression, multiplication and division are computed next from left to right, followed by addition and subtraction.

16. A: 847.90.

The hundredths place value is located two digits to the right of the decimal point (the digit 9 in the original number). The digit to the right of the place value is examined to decide whether to round up or keep the digit. In this case, the digit 6 is 5 or greater so the hundredth place is rounded up. When rounding up, if the digit to be increased is a 9, the digit to its left is increased by one and the digit in the desired place value is made a zero. Therefore, the number is rounded to 847.90.

17. C: Perimeter is found by calculating the sum of all sides of the polygon.

$9 + 9 + 9 + 8 + 8 + s = 56$, where s is the missing side length. Therefore, 43 plus the missing side length is equal to 56. The missing side length is 13 cm.

18. A: $16\frac{1}{2}$. A mixed number contains both a whole number and either a fraction or a decimal. Therefore, the mixed number is $16\frac{1}{2}$.

19. E: $9\frac{3}{10}$. To convert a decimal to a fraction, remember that any number to the left of the decimal point will be a whole number. Then, since 0.3 goes to the tenths place, it can be placed over 10.

20. C: To solve for the value of b, both sides of the equation need to be equalized.

Start by cancelling out the lower value of -4 by adding 4 to both sides:

$$5b - 4 = 2b + 17$$
$$5b - 4 + 4 = 2b + 17 + 4$$
$$5b = 2b + 21$$

The variable b is the same on each side, so subtract the lower 2b from each side:

$$5b = 2b + 21$$
$$5b - 2b = 2b + 21 - 2b$$
$$3b = 21$$

Then divide both sides by 3 to get the value of b:

$$3b = 21$$

$$\frac{3b}{3} = \frac{21}{3}$$

$$b = 7$$

21. A: 63.96

This question involves the percent formula. Since we're beginning with a percent, also known as a number over 100, we'll put 39 on the right side of the equation:

$$\frac{x}{164} = \frac{39}{100}$$

Now, multiply 164 and 39 to get 6,396, which then needs to be divided by 100.

$$6,396 \div 100 = 63.96$$

22. C: $\frac{1}{3}$ of the shirts sold were patterned. Therefore, $1 - \frac{1}{3} = \frac{2}{3}$ of the shirts sold were solid. Anytime "of" a quantity appears in a word problem, multiplication needs to be used. Therefore, $192 \times \frac{2}{3} =$ $192 \times \frac{2}{3} = \frac{384}{3} = 128$ solid shirts were sold. The entire expression is $192 \times \left(1 - \frac{1}{3}\right)$.

23. B: A rectangle is a specific type of parallelogram. It has 4 right angles. A square is a rhombus that has 4 right angles. Therefore, a square is always a rectangle because it has two sets of parallel lines and 4 right angles.

24. E: Recall the formula for area, area $=$ length $\times$ width. The answer must be in square inches, so all values must be converted to inches. Half of a foot is equal to 6 inches. Therefore, the area of the rectangle is equal to:

$$6 \text{ in} \times \frac{11}{2} \text{ in} = \frac{66}{2} \text{ in}^2 = 33 \text{ in}^2$$

25. D: 80 percent. To convert a fraction to a percent, the fraction is first converted to a decimal. To do so, the numerator is divided by the denominator: $4 \div 5 = 0.8$. To convert a decimal to a percent, the number is multiplied by 100: $0.8 \times 10 = 80\%$.

26. C: 80 min. To solve the problem, a proportion is written consisting of ratios comparing distance and time. One way to set up the proportion is:

$$\frac{3}{48} = \frac{5}{x} \left(\frac{distance}{time} = \frac{distance}{time}\right)$$

where x represents the unknown value of time. To solve a proportion, the ratios are cross-multiplied:

$$(3)(x) = (5)(48) \rightarrow 3x = 240$$

The equation is solved by isolating the variable, or dividing by 3 on both sides, to produce $x = 80$.

27. A: Every 8 ml of medicine requires 5 ml. The 45 ml first needs to be split into portions of 8 ml. This results in $\frac{45}{8}$ portions. Each portion requires 5 ml. Therefore:

$$\frac{45}{8} \times 5 = 45 \times \frac{5}{8} = \frac{225}{8}$$

ml is necessary.

28. C: Volume of this 3-dimensional figure is calculated using length x width x height. Each measure of length is in inches. Therefore, the answer would be labeled in cubic inches.

29. A: A common denominator must be found. The least common denominator is 15 because it has both 5 and 3 as factors. The fractions must be rewritten using 15 as the denominator.

30. C: A compass is a tool that can be used to draw a circle. The compass would be drawn by using the length of the radius, which is half of the diameter.

31. D: A dollar contains 20 nickels. Therefore, if there are 12 dollars' worth of nickels, there are $12 \times 20 = 240$ nickels.

Each nickel weighs 5 grams. Therefore, the weight of the nickels is $240 \times 5 = 1,200$ grams.

Adding in the weight of the empty piggy bank, the filled bank weighs 2,250 grams.

32. D: To find Denver's total snowfall, 3 must be multiplied times $27\frac{3}{4}$. In order to easily do this, the mixed number should be converted into an improper fraction.

$$27\frac{3}{4} = \frac{27 \times 4 + 3}{4} = \frac{111}{4}$$

Therefore, Denver had approximately $\frac{3 \times 111}{4} = \frac{333}{4}$ inches of snow. The improper fraction can be converted back into a mixed number through division.

$$\frac{333}{4} = 83\frac{1}{4} \text{ inches}$$

33. B: Ordinal numbers represent a ranking. Placing second in a competition is a ranking among the other participants of the spelling bee.

34. D: 13,078. The power of 10 by which a digit is multiplied corresponds with the number of zeros following the digit when expressing its value in standard form. Therefore:

$$(1 \times 10^4) + (3 \times 10^3) + (7 \times 10^1) + (8 \times 10^0)$$

$$10,000 + 3,000 + 70 + 8 = 13,078$$

35. C: 374.04. The formula for finding the area of a regular polygon is $A = \frac{1}{2} \times a \times P$ where a is the length of the apothem (from the center to any side at a right angle), and P is the perimeter of the figure. The apothem a is given as 10.39, and the perimeter can be found by multiplying the length of one side by the number of sides (since the polygon is regular): $P = 12 \times 6 \rightarrow P = 72$. To find the area, substitute the values for a and P into the formula

$$A = \frac{1}{2} \times a \times P$$

$$A = \frac{1}{2} \times (10.39) \times (72) \rightarrow A = 374.04$$

36. E: 216 cm². Because area is a two-dimensional measurement, the dimensions are multiplied by a scale that is squared to determine the scale of the corresponding areas. The dimensions of the rectangle are multiplied by a scale of 3. Therefore, the area is multiplied by a scale of 3^2 (which is equal to 9):

$$24 \ cm^2 \times 9 = 216 \ cm^2$$

37. B: A number is prime because its only factors are itself and 1. Positive numbers (greater than one) can be prime numbers.

38. C: $\frac{31}{36}$

Set up the problem and find a common denominator for both fractions.

$$\frac{5}{12} + \frac{4}{9}$$

Multiply each fraction across by 1 to convert to a common denominator.

$$\frac{5}{12} \times \frac{3}{3} + \frac{4}{9} \times \frac{4}{4}$$

Once over the same denominator, add across the top. The total is over the common denominator.

$$\frac{15 + 16}{36} = \frac{31}{36}$$

39. A: Order of operations follows PEMDAS—Parentheses, Exponents, Multiplication and Division from left to right, and Addition and Subtraction from left to right.

40. C: 0.63

Divide 5 by 8, which results in 0.63.

41. D: 8,685

Set up the problem, with the larger number on top. Begin subtracting with the far-right column (ones). Borrow 10 from the column to the left, when necessary.

42. A: The place value to the right of the thousandth place, which would be the ten-thousandth place, is what gets used. The value in the thousandth place is 7. The number in the place value to its right is greater than 4, so the 7 gets bumped up to 8. Everything to its right turns to a zero, to get 245.2680. The zero is dropped because it is part of the decimal.

43. D: 37.797

Set up the problem, larger number on top and numbers lined up at the decimal. Begin subtracting with the far-right column. Borrow 10 from the column to the left, when necessary.

44. C: Numbers should be lined up by decimal places before subtraction is performed. This is because subtraction is performed within each place value. The other operations, such as multiplication, division, and exponents (which is a form of multiplication), involve ignoring the decimal places at first and then including them at the end.

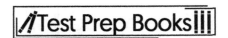

45. C: $\frac{19}{24}$

Set up the problem and find a common denominator for both fractions.

$$\frac{23}{24} - \frac{1}{6}$$

Multiply each fraction across by 1 to convert to a common denominator.

$$\frac{23}{24} \times \frac{1}{1} - \frac{1}{6} \times \frac{4}{4}$$

Once over the same denominator, subtract across the top.

$$\frac{23 - 4}{24} = \frac{19}{24}$$

46. E: $\frac{2}{9}$

Set up the problem and find a common denominator for both fractions.

$$\frac{43}{45} - \frac{11}{15}$$

Multiply each fraction across by 1 to convert to a common denominator.

$$\frac{43}{45} \times \frac{1}{1} - \frac{11}{15} \times \frac{3}{3}$$

Once over the same denominator, subtract across the top.

$$\frac{43 - 33}{45} = \frac{10}{45}$$

Reduce.

$$\frac{10 \div 5}{45 \div 5} = \frac{2}{9}$$

47. B: $\frac{14}{25}$

Since 0.56 goes to the hundredths place, it can be placed over 100:

$$\frac{56}{100}$$

Essentially, the way we got there is by multiplying the numerator and denominator by 100:

$$\frac{0.56}{1} \times \frac{100}{100} = \frac{56}{100}$$

Then, the fraction can be simplified down to $\frac{14}{25}$:

$$\frac{56}{100} \div \frac{4}{4} = \frac{14}{25}$$

48. A: 2,504,774

Line up the numbers (the number with the most digits on top) to multiply. Begin with the right column on top and the right column on bottom.

Move one column left on top and multiply by the far-right column on the bottom. Remember to add the carry over after you multiply. Continue that pattern for each of the numbers on the top row.

Starting on the far-right column on top repeat this pattern for the next number left on the bottom. Write the answers below the first line of answers; remember to begin with a zero placeholder. Continue for each number in the top row.

Starting on the far-right column on top, repeat this pattern for the next number left on the bottom. Write the answers below the first line of answers. Remember to begin with zero placeholders.

Once completed, ensure the answer rows are lined up correctly, then add.

49. A: The first step is to divide up $150 into four equal parts. $\frac{150}{4}$ is 37.5, so she needs to save an average of $37.50 per day.

50. D: The teacher would be introducing fractions. If a pie was cut into 6 pieces, each piece would represent $\frac{1}{6}$ of the pie. If one piece was taken away, $\frac{5}{6}$ of the pie would be left over.

51. A: $\frac{810}{2921}$

Line up the fractions.

$$\frac{15}{23} \times \frac{54}{127}$$

Multiply across the top and across the bottom:

$$\frac{15 \times 54}{23 \times 127} = \frac{810}{2921}$$

52. D: There are no millions, so the millions period consists of all zeros. 182 is in the billions period, 36 is in the thousands period, 421 is in the hundreds period, and 356 is the decimal.

53. D: $11\frac{3}{4}$

Set up the division problem.

$$1 \quad 6 \overline{\big)\ 1 \quad 8 \quad 8}$$

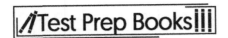

16 does not go into 1 but does go into 18 so start there.

$$
\begin{array}{r}
\quad 1\ \ \ 1 \\
1\quad 6\,\overline{)1\ \ 8\ \ 8} \\
-\ \ \ 1\ \ 6 \\
\hline
2\ \ 8 \\
-\ \ 1\ \ 6 \\
\hline
1\ \ 2
\end{array}
$$

The result is $11\frac{12}{16}$

Reduce the fraction for the final answer.

$$11\frac{3}{4}$$

54. D: Division can be computed as a repetition of subtraction problems by subtracting multiples of 24.

55. E: The number of days can be found by taking the total amount Bernard needs to make and dividing it by the amount he earns per day:

$$\frac{300}{80} = \frac{30}{8} = \frac{15}{4} = 3.75$$

But Bernard is only working full days, so he will need to work 4 days, since 3 days is not a sufficient amount of time.

56. E: The value went up by:

$$\$165,000 - \$150,000 = \$15,000$$

Out of $150,000, this is $\frac{15,000}{150,000} = \frac{1}{10}$. Convert this to having a denominator of 100, the result is $\frac{10}{100}$ or 10%.

57. B: A number raised to an exponent is a compressed form of multiplication.

For example, $10^3 = 10 \times 10 \times 10$.

58. B: The perimeter of a rectangle is the sum of all four sides.

Therefore, the answer is:

$$P = 14 + 8\frac{1}{2} + 14 + 8\frac{1}{2}$$

$$14 + 14 + 8 + \frac{1}{2} + 8 + \frac{1}{2} = 45 \text{ square inches}$$

59. C: Kilograms, grams, kilometers, and meters. Inches, pounds, and baking measurements, such as tablespoons, are not part of the metric system. Kilograms, grams, kilometers, and meters are part of the metric system.

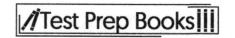

60. D: It shows the associative property of multiplication. The order of multiplication does not matter, and the grouping symbols do not change the final result once the expression is evaluated.

Reading

1. B: Mr. Button's wife is about to have a baby. The passage begins by giving the reader information about traditional birthing situations. Then, we are told that Mr. and Mrs. Button decide to go against tradition to have their baby in a hospital. The next few passages are dedicated to letting the reader know how Mr. Button dresses and goes to the hospital to welcome his new baby. There is a doctor in this excerpt, as Choice *C* indicates, and Mr. Button does put on clothes, as Choice *D* indicates. However, Mr. Button is not going to the doctor's office nor is he about to go shopping for new clothes.

2. A: The tone of the above passage is nervous and excited. We are told in the fourth paragraph that Mr. Button "arose nervously." We also see him running without caution to the doctor to find out about his wife and baby—this indicates his excitement. We also see him stuttering in a nervous yet excited fashion as he asks the doctor if it's a boy or girl. Though the doctor may seem a bit abrupt at the end, indicating a bit of anger or shame, neither of these choices is the overwhelming tone of the entire passage. Despite the circumstances, joy and gratitude are not the main tone in the passage.

3. C: Dedicated. Mr. Button is dedicated to the task before him. Choice *A*, numbed, Choice *B*, chained, and Choice *D*, moved, all could grammatically fit in the sentence. However, they are not synonyms with *consecrated* like Choice *C* is.

4. D: Giving readers a visual picture of what the doctor is doing. The author describes a visual image— the doctor rubbing his hands together—first and foremost. The author may be trying to make a comment about the profession; however, the author does not "explain the detail of the doctor's profession" as Choice *B* suggests.

5. D: To introduce the setting of the story and its characters. We know we are being introduced to the setting because we are given the year in the very first paragraph along with the season: "one day in the summer of 1860." This is a classic structure of an introduction of the setting. We are also getting a long explanation of Mr. Button, what his work is, who is related to him, and what his life is like in the third paragraph.

6. B: "Talk sense!" is an example of an imperative sentence. An imperative sentence gives a command. The doctor is commanding Mr. Button to talk sense. Choice *A* is an example of an exclamatory sentence, which expresses excitement. Choice *C* is an example of an interrogative sentence—these types of sentences ask questions. Choice *D* is an example of a declarative sentence. This means that the character is simply making a statement.

7. B: The word *anachronism* most nearly means misplacement. Choice *A*, comparison, is an analogy or similarity to something. Choice *C*, *aberration*, means abnormality. Choice *D*, *amelioration*, means improvement.

8. C: This passage can best be described as a narrative, which is a type of passage that tells a story. Choice *A*, expository, is a text organized logically to investigate a problem. Choice *B*, descriptive, is a text that mostly goes about describing something or someone in detail. Choice *D*, persuasive, is a text that is organized as an argument and meant to persuade the audience to do something.

9. A: To explain the setting of the narrative and give information about the story. The setting of a narrative is the time and place. We see from the first paragraph that the year is 1860. We also can discern that it is summer, and Mr. and Mrs. Button are about to have a baby. This tells us both the setting and information about the story.

10. D: Something is unusual about the birth of the baby. The word "curious" is thrown in at the end twice, which tells us the doctor is suspicious about something having to do with the birth of the baby, since that is the most recent event to happen. Mr. Button is acting like a father who is expecting a baby, and the doctor seems confused about something.

11. C: This is a tricky question, but it can be solved through careful context analysis and vocabulary knowledge. One can infer that the use of "expedient," while not necessarily very positive, isn't inherently bad in this context either. Note how in the next line, he says, "but most governments are usually, and all governments are sometimes, inexpedient." This use of "inexpedient" indicates that a government becomes a hindrance rather than a solution; it slows progress rather than helps facilitate progress. Thus, Choice *A* and Choice *D* can be ruled out because these are more of the result of government, not the intention or initial design. Choice *B* makes no logical sense. Therefore, Choice *C* is the best description of *expedient.* Essentially, Thoreau is saying that government is constructed as a way of developing order and people's rights, but the rigidness of government soon inhibits justice and human rights.

12. B: While Choice *D* is the only answer that mentions the Mexican War directly, Thoreau clearly thinks the war is unnecessary because the people generally didn't consent to the war. Choices *A*, *B*, and *C* are all correct to a degree, but the answer asks for the best description. Therefore, Choice *B* is the most accurate representation of Thoreau's views. Essentially, Thoreau brings to light the fact that the few people in power can twist government and policy for their own needs.

13. D: Choices *A* and *B* are completely incorrect. Thoreau is not defending government in any way. His views are set against government. As mentioned in the text, he appreciates little government but favors having no government structure at all. The text is reflective by nature but not reflective on the stages of government abuses as Choice *C* suggests. Choice *D* is the more appropriate answer because of the presence of evidence in the text. Thoreau cites current events and uses them to illustrate the point he's trying to make.

14. C: One of Thoreau's biggest criticisms of government is its capacity to impose on the people's freedoms and liberties, enacting rules that the people don't want and removing power from the individual. None of the scenarios directly impose specific regulations or restrictions on the people, except Prohibition. Prohibition removed the choice to consume alcohol in favor of abstinence, which was favored by the religious conservatives of the time. Thus, Thoreau would point out that this is a clear violation of free choice and an example of government meddling.

15. A: Choice *B* is totally irrelevant. Choice *C* is also incorrect; Thoreau never personifies government. Also, this doesn't coincide with his wooden gun analogy. Choice *D* is compelling because of its language but doesn't define the statement. Choice *A* is the most accurate summary of the main point of Thoreau's statement.

16. B: Thoreau specifically cites that legislators "are continually putting in their way." This reflects his suspicion and concern of government intervention. Recall that Thoreau continually mentions that government, while meant as a way to establish freedom, is easily used to suppress freedom, piling on

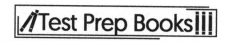

regulations and rules that inhibit progress. Choice *B* is the answer that most directly states how Thoreau sees government getting in the way of freedom.

17. D: This passage is an expository essay, which means that an idea is investigated and then expanded upon. Thoreau is investigating the idea of government here and how the U.S. government works in relation to the people.

18. C: The word *liable* most nearly means susceptible. The text says, "The government itself, which is only the mode which the people have chosen to execute their will, is equally liable to be abused and perverted before the people can act through it." Thoreau is saying here that the government is vulnerable enough to be abused. Choice *A*, *paramount*, means having importance. Choice *B*, *inconceivable*, means unbelievable. Choice *D*, *detrimental*, means damaging.

19. B: Thoreau is talking about the American government. We see this information in the beginning of the second paragraph. The passage mentions the Mexican War, but the passage itself does not relay the vulnerabilities of the Mexican government. The other two countries are not mentioned in the passage.

20. B: Posterity means all future generations of people. The sentence would say, "The American government—what is it but a tradition, though a recent one, endeavoring to transmit itself unimpaired to all future generations of people . . ." Choice *A* would be the definition of *royalty*. Choice *C* would be the definition of an *educator*. Choice *D* would be the definition of a *philosopher*.

21. A: Choice *D* can be eliminated because the Salem witch trials aren't mentioned. While sympathetic to the plight of the accused, the author doesn't demand or urge the reader to demand reparations to the descendants; therefore, Choice *B* can also be ruled out. It's clear that the author's main goal is to educate the reader and shed light on the facts and hidden details behind the case. However, his focus isn't on the occult, but the specific Lancashire case itself. He goes into detail about suspects' histories and ties to Catholicism, revealing how the fears of the English people at the time sealed the fate of the accused witches. Choice *A* is correct.

22. B: It's important to note that these terms may not be an exact analog for *enduring*. However, through knowledge of the definition of *enduring*, as well as the context in which it's used, an appropriate synonym can be found. Plugging "circumstantial" into the passage in place of "enduring" doesn't make sense. Nor does "un-original" work, since this particular case of witchcraft stands out in history. "Wicked" is very descriptive, but this is an attribute applied to people, not events; therefore, this is an inappropriate choice as well. *Enduring* literally means long lasting, referring to the continued interest in this particular case of witchcraft. Therefore, it's a popular topic of 1600s witch trials, making "popular," Choice *B*, the best choice.

23. D: Choices *A* and *B* are irrelevant and incorrect. The use of quotes lends credibility to the author. However, the presence of quotes alone doesn't necessarily mean that the author has a qualified perspective. What establishes the writer as a reliable voice is that the author's previous writing on the subject has been published before. This qualification greatly establishes the author's credentials as a historical writer, making Choice *D* the correct answer.

24. B: Choice *B* is the best answer because it ultimately encompasses the potentiality of Choices *C* and *D*. Choice *A* is incorrect because it takes the statement literally. For Choice *C*, it's possible that evidence was tampered with or even falsified, but this statement doesn't refer to this. While the author alludes that there may have been evidence tampering and potentially corruption (Choice *D*), what the writer is directly saying is that the documentation of the court indicates an elaborate trial.

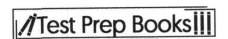

25. C: Several of these answers could have contributed to the fear and political motivations around the Lancashire witch trials. What this answer's looking for is very specific: political motivations and issues that played a major role in the case. Choice *C* clearly outlines the public fears of the time. It also describes how the government can use this fear to weed out and eliminate traces of Catholicism (and witchcraft too). Catholicism and witchcraft were seen as dangerous and undermining to English Protestantism and governance. Choice *D* can be eliminated; while this information may have some truth and is certainly consistent with the general fear of witchcraft, the details about Lancashire's ancient history aren't mentioned in the text. Choice *A* is true but not necessarily political in nature. Choice *B* is very promising, though not outright mentioned.

26. D: The best evidence comes from Alizon herself. The text mentions that she confessed to bewitching John Law, thinking that she did him harm. From here she names her grandmother, who she believes corrupted her. Choice *B* can be ruled out; spectral evidence isn't mentioned. The case draws on knowledge of superstition of witchcraft, but this in itself can't be considered evidence, so Choice *A* is incorrect. Choice *C* isn't evidence in a modern sense; rumors have no weight in court and therefore are not evidence. While this is used as evidence to some degree, this still isn't the *best* evidence against Alizon and the witches.

27. B: This type of passage would be considered expository, which is an informative passage. Choice *A*, persuasive, means to take a side of an argument, and this essay is merely divulging information. Choice *C*, narrative, means to tell a story. Although a story is being told indirectly, the essay doesn't follow a traditional narrative. Choice *D*, descriptive, means a detailed description of a person or place.

28. B: According to the passage, 20 people were arrested as witches in the Lancashire trials. The essay tells us that "ten were hanged, one died in goal (jail), one was sentenced to death in the pillory, and eight were acquitted."

29. D: The word *commemorative* means in honor of something. The context clue here includes the "works of fiction" by the authors Blake Morrison, Ann Duffy, and Jeanette Winterson, no doubt to celebrate the preserved history of the famous trial.

30. C: Malkin tower is the house of Alizon's grandmother. It is also a place where a meeting of witches was said to be held in the passage. The passage says, "The net was widened still further at the end of April when Alizon's younger brother James and younger sister Jennet, only nine years old, came up between them with a story about a "great meeting of witches" at their grandmother's house, known as Malkin Tower."

31. A: It introduces certain insects that transition from water to air. Choice *B* is incorrect because although the passage talks about gills, it is not the central idea of the passage. Choices *C* and *D* are incorrect because the passage does not "define" or "invite," but only serves as an introduction to stoneflies, dragonflies, and mayflies and their transition from water to air.

32. C: The act of shedding part or all of the outer shell. Choices *A*, *B*, and *D* are incorrect. The word in the passage is mentioned here: "But it is advisable in connection with our present subject to dwell especially on some insects that remain continually under water till they are ready to undergo their final moult and attain the winged state, which they pass entirely in the air."

33. B: The first paragraph serves as a contrast to the second. Notice how the first paragraph goes into detail describing how insects are able to breathe air. The second paragraph acts as a contrast to the first by stating "[i]t is of great interest to find that, nevertheless, a number of insects spend much of their

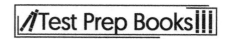

time under water." Watch for transition words such as "nevertheless" to help find what type of passage you're dealing with.

34: C: The stage of preparation in between molting is acted out in the water, while the last stage is in the air. Choices *A, B,* and *D* are all incorrect. *Instars* is the phase between two periods of molting, and the text explains when these transitions occur.

35. C: The author's tone is informative and exhibits interest in the subject of the study. Overall, the author presents us with information on the subject. One moment where personal interest is depicted is when the author states, "It is of great interest to find that, nevertheless, a number of insects spend much of their time under water."

36. C: Their larva can breathe the air dissolved in water through gills of some kind. This is stated in the last paragraph. Choice *A* is incorrect because the text mentions this in a general way at the beginning of the passage concerning "insects as a whole." Choice *B* is incorrect because this is stated of beetles and water-bugs, and not the insects in question. Choice *D* is incorrect because this is the opposite of what the text says of instars.

37. B: According to the passage, boatmen and scorpions have some way of protecting their spiracles when submerged. We see this in the second paragraph, which says "(boatmen and scorpions) which have some way of protecting their spiracles when submerged."

38. A: The best answer Choice is *A*: the author believes that the stonefly, dragonfly, and mayfly larvae are better prepared to live beneath the water because they have gills that allow them to do so. We see this when the author says "But the larva of a stone-fly, a dragon-fly, or a may-fly is adapted more completely than these for aquatic life; it can, by means of gills of some kind, breathe the air dissolved in water."

39. C: Because a vast majority of insects have wings and also have the ability to breathe underwater. The entire first paragraph talks of how insects have wings, and how insects also have "a system of branching air-tubes" that carries oxygen to the insect's tissues.

40. B: The word *preeminently* most nearly means *above all* or *in particular*. The author is saying that above all, insects are creatures of both land and water.

Science

1. A: One cycle takes 300 months. It starts with the population being low, rising and reaching a peak, and then falling again. It takes 200 months to reach the peak, Choice *B*. Three cycles could be completed in 800 months, Choice *C*, and the second cycle would have started by 400 months, Choice *D*.

2. B: Looking at Figure 2, the fox is the predator. When the diagram notes that the predator population decreases on the left side, there is only one fox left. As it increases, as noted on the right side, there are two foxes drawn. Foxes are also much larger than rabbits and would be able to catch them much easier than the other way around.

3. D: When the prey population decreases, the predators have less food, i.e. prey, to feed on. This causes the predator population to dwindle. An increase in the prey population, Choice *A*, would actually increase the predator population because they would have more food, which would lengthen survival and increase reproduction. Seasons do not affect the predator population in this situation, Choice *B*.

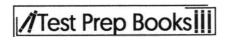

Generally, prey do not have the ability to attack their predators, Choice *C*, due to physical constraints, such as differences in size.

4. A: When the predator population decreases, the rate of survival of the prey population increases, and they can then also reproduce more. An increase in the predator population, Choice *B*, would cause the prey population to decrease. Weather and amount of sunlight, Choice *C*, does not affect the growth of the prey population. The prey population is cyclical and does not remain the same size, Choice *D*.

5. C: Prey populations can develop different features to try and hide from and escape the predator population. The features help them blend into their environment, such as Choice *B*, or help them identify predators early and quickly, Choices *A* and *D*. Choice *C* would just allow the predators to hear the prey easily.

6. B: Sunlight comes into the greenhouse as short-wavelength IR. As it is absorbed by surfaces in the greenhouse, it is converted to long-wavelength IR. The long-wavelength IR gets trapped inside the greenhouse and bounces off the surfaces and glass and remains inside the greenhouse. Since short-wavelength IR can enter the greenhouse, it also has the ability to leave the greenhouse, making Choice *A* incorrect. The greenhouse feels warmer, not cooler, than outside, so Choice *C* is incorrect. Water is not involved in the reaction noted in Figure 1, so Choice *D* is also incorrect.

7. D: Looking at the graph in Figure 2, the greenhouse plant grew taller than the outside plant. The bars representing the greenhouse plant are taller at 3 time points that Scientist A measured. Greenhouses trap sunlight, warm air, and gases, such as CO_2 inside the greenhouse, so plants have an increased rate of photosynthesis, allowing them to grow faster. The plants are also protected from pests inside the greenhouse.

8. B: Sunlight enters the greenhouse as short-wavelength IR and get converted to long-wavelength IR. This process also gives off heat and makes the greenhouse feel warmer than the outside climate. Water and oxygen, Choices *A* and *D*, are not involved in this reaction. The plants remain the same and do not get converted into anything else, Choice *C*.

9. A: Short-wavelength IR enters the greenhouse in the form of sunlight. It can pass easily through the glass and can therefore pass easily back out to the outside environment. The long-wavelength IR, Choice *D*, gets trapped inside the greenhouse.

10. C: The plants inside a greenhouse are protected from many pests that can be found in the outside environment. The air is warmer in the greenhouse, so Choice *A* is incorrect. More photosynthesis occurs because of the increased sunlight energy that stays in the greenhouse, making Choices *B* and *D* incorrect.

11. D: The equivalence point occurs when just enough of the unknown solution is added to completely neutralize the known solution. In Figure 1, at the halfway point of the curve, the equivalence point is when 25 volumes of NaOH have been added to the solution. The pH is 7 at this point also, which is a neutralization of the HCl, strong acid.

12. C: The acid dissociation constant is the pK_a of the solution. It is found at the halfway point between the beginning of the curve and the equivalence point, where the solution would have a pH of 7 and be completely neutralized. In Figure 3, it is marked as 4.76.

13. A: Looking at Figure 3, the vertical axis on the left side has information about the pH of the solution. The horizontal axis at the bottom has information about how much basic solution containing OH⁻ is being added to the acetic acid. When the OH⁻ is at 0, and none has been added yet, the pH of the acetic acid is marked as 2.

14. B: Looking at the chemical equation in Figure 2, the reactants are on the left side and the products are on the right side. HCl and NaOH, Choices *A* and *C*, are the reactants of the equation. NaCl is the salt that is formed as one of the products of the reaction. The chloride ion, Choice *D*, is not formed in this reaction.

15. A: The equivalence point occurs in all titration reactions when the solution is neutralized. If an acid and base are being titrated, the solution is no longer acidic or basic, Choices *B* and *C*. It reaches a pH of 7 and is considered neutral.

16. A: Oxygen-poor blood is pumped to the lungs before returning to the heart. Oxygen is transferred from the airways of the lungs into the blood. The blood becomes rich with oxygen and then returns to the heart so that it can bring oxygen and nutrients to other organs of the body, such as the brain, stomach, and kidney, Choices *B*, *C*, and *D*.

17. C: The mitral valve has two folds. The tricuspid valve, Choice *B*, has three folds. The valve between the left atrium and ventricle has two valves, as was noted in the descriptive passage. Correlating this information to Figure 1, the name of the valve between these two chambers is the mitral valve. Choices *A* and *D* are vessels that carry blood through them and are not names of valves.

18. D: Arteries carry oxygen-rich blood away from the heart to the rest of the body. The aorta is the largest artery in the body. Veins, Choice *A*, carry oxygen-poor blood to the heart and lungs. Airway, Choice *C*, is found in the respiratory system and carries air in and out of the body.

19. C: A normal heartrate for a child is between 90 and 120 beats per minute. Choice *C*, 100 beats per minute, falls within this range. Choices *A* and *D* are within the normal range for an adult but not for a child. Children's hearts pump blood faster than adults' hearts through the body. Choice *B*, 125, is higher than the normal range for children.

20. B: Looking at Figure 1 and following the red arrow in the aorta backwards, the left ventricle is where the blood is coming from directly before it enters the aorta. Blood flows from the atria to the ventricles, so it enters the left atrium before the left ventricle and then the aorta, Choice *D*. Oxygen-poor blood is on the right side of the heart and flows from the right atrium to the right ventricle before flowing to the lungs to get re-oxygenated, Choices *A* and *C*.

21. C: Lava is a type of molten rock. When molten rock cools down and hardens, it forms igneous rock. Sedimentary and metamorphic rocks, Choices *A* and *B*, are not formed from molten rock. Choice *D* is incorrect because lava does cool down eventually and becomes hard.

22: A: Looking at Figure 1, the granite is found in column A. The description of the rocks in these columns says that these rocks were formed from molten rock. When molten rock cools, it forms igneous rock. Column C describes how metamorphic rocks are formed, Choice *B*. Column B describes how sedimentary rocks are formed, Choice *C*. Fossils, Choice *D*, are not rocks but are formed into sedimentary rock.

23: D: Sedimentary rocks are formed from soft materials, such as sand, shells, and pebbles. This allows for fossils to form because the remains of animals or plants can be pressed into the softer rock material and leave their imprint. Fossils cannot form in igneous, Choice *A*, or metamorphic, Choice *C*, rocks. Bones are something that can actually make a fossil imprint, Choice *B*.

24. B: Rocks that are formed by changes in heat and pressure are called metamorphic rocks, which is how the rocks in Column C are described. Slate is found in Column C of Figure 1. Sandstone and limestone, Choices *A* and *D*, are both found in Column B, which describes sedimentary rock. Granite, Choice *B*, is found in Column A, which describes sedimentary rock.

25. A: Sedimentary rock is formed from sand, shells, and pebbles, all of which are found in abundance at the beach. Shells, Choice *B*, are something that contribute to the formation of sedimentary rock. Igneous rock, Choice *C*, is formed from molten rock, which would likely be much too hot to be found on most beaches. The surface environment of a beach likely does not undergo changes in heat and pressure enough to form metamorphic rock, Choice *D*.

26. D: Helium gas is not one of the major direct or indirect greenhouse gases. Looking at Figure 2, Choices *A*, *B*, and *C* are part of the pie chart as greenhouse gases found in the atmosphere.

27. B: The two major wedges of the pie chart in Figure 3 are Transportation and Electricity. Using a car, Choice *C*, produces a lot of carbon dioxide. Walking instead of driving a car would not produce any carbon dioxide. Choices *A* and *D* use electricity and leaving either the lights or the television on would need a constant source of electricity, producing lots of carbon dioxide.

28. A: Looking at Figure 2, carbon dioxide, Choice *D*, takes up the largest wedge of the pie chart at 54.7%. The next largest wedge is methane at 30%. Nitrous oxide, Choice *C*, takes up only 4.9% and fluorinated gases, Choice *B*, takes up 0.6%.

29. C: Looking at Figure 1, in the right picture where there are more greenhouse gases, the re-emitted heat arrow is larger as more heat gets trapped in the Earth's atmosphere. Space and the Sun, Choices *A* and *B*, remain outside the Earth's atmosphere. Water vapor, Choice *D*, is not a part of the diagrams for the greenhouse effect.

30. D: Figures 2 and 3 are pie charts. Circular charts that are broken up into wedges, or pie pieces, are called pie charts. Scatter plots, Choice *A*, have specific point markers to mark each data point. In line graphs, Choice *B*, data are represented by connecting lines. In bar graphs, Choice *C*, data are represented by vertical or horizontal bars.

31. A: An anemometer measures air speed. Wind is the movement of air, so an anemometer would be able to measure wind speed. A barometer, Choice *B*, measures air pressure. A thermometer, Choice *C*, measures temperature. A wind vane, Choice *D*, shows which direction the wind is blowing.

32. B: Figure 1 is a graph showing the temperature on different days. Temperature is measured using a thermometer. A rain gauge, Choice *A*, would allow a meteorologist to record amounts of rainfall. A barometer, Choice *C*, measures air pressure. An anemometer, Choice *D*, measures air speed.

33. C: A wind vane shows which direction the wind is blowing. If it is pointing north, the wind is blowing in a northern direction. It would not be blowing in an eastern direction, Choice *A*, or a southern direction, Choice *D*, since that is the opposite direction of north. A wind vane simply tells wind direction and does not determine whether a storm is coming, Choice *B*.

34. A: Looking at the line graph in Figure 2, the lowest point is marked for January, with approximately 2 mm of rainfall. April, Choice *B*, has approximately 23 mm rainfall, September, Choice *C*, has 10 mm of rainfall, and October, Choice *D*, has the highest rainfall at 60 mm.

35. D: Reading the graph in Figure 2, at June, the rainfall is approximately 20 mm. The trend line marks all of the data collected for each month. For June, the trend line is just about at the 20 mm mark from the vertical axis on the left side of the graph.

36. B: Plants use chloroplasts to turn light energy into glucose. Animal cells do not have this ability. Comparing Figures 1 and 2, chloroplasts can be found in the plant cell but not the animal cell.

37. C: The nucleolus is always located inside the nucleus. It contains important hereditary information about the cell that is critical for the reproductive process. Chloroplasts, Choice *A*, are only located in plant cells. It is not found in the mitochondria, Choice *B*, or attached to the cell membrane, Choice *D*.

38. A: Looking at the table in Figure 3, each organelle is described, and mitochondria is described as the powerhouse of the cell. The nucleus, Choice *B*, contains the cell's DNA. The ribosomes, Choice *C*, build proteins. The cell wall, Choice *D*, maintains the shape of plant cells and protects its contents.

39. D: Figure 3 describes the functions of the organelles. The cell membrane surrounds the cell and regulates which molecules can move in and out of the cell. Ribosomes build proteins, Choice *A*. Lysosomes, Choice *B*, break down large molecules. The nucleus, Choice *C*, contains the cell's DNA.

40. C: Figure 3 describes the functions of the organelles. Chloroplasts are responsible for photosynthesis in plant cells, which is the process of converting sunlight energy to glucose energy. The cell wall helps maintain the cell's shape, Choice *A*. The nucleus contains the cell's DNA, Choice *B*. Ribosomes build proteins, Choice *D*.

ACT Practice Test #3

English

Questions 1–15 are based on the following passage:

The knowledge of an aircraft engineer is acquired through years of education, and special (1) licenses are required. Ideally, an individual will begin his or her preparation for the profession in high school (2) by taking chemistry physics trigonometry and calculus. Such curricula will aid in (3) one's pursuit of a bachelor's degree in aircraft engineering, which requires several physical and life sciences, mathematics, and design courses.

(4) Some of universities provide internship or apprentice opportunities for the students enrolled in aircraft engineer programs. A bachelor's in aircraft engineering is commonly accompanied by a master's degree in advanced engineering or business administration. Such advanced degrees enable an individual to position himself or herself for executive, faculty, and/or research opportunities. (5) These advanced offices oftentimes require a Professional Engineering (PE) license which can be obtained through additional college courses, professional experience, and acceptable scores on the Fundamentals of Engineering (FE) and Professional Engineering (PE) standardized assessments.

(6) Once the job begins, this lines of work requires critical thinking, business skills, problem solving, and creativity. This level of (7) expertise (8) allows aircraft engineers to (9) apply mathematical equation and scientific processes to aeronautical and aerospace issues or inventions. (10) For example, aircraft engineers may test, design, and construct flying vessels such as airplanes, space shuttles, and missile weapons. As a result, aircraft engineers are compensated with generous salaries. In fact, in May 2014, the lowest 10 percent of all American aircraft engineers earned less than $60,110 while the highest paid ten-percent of all American aircraft engineers earned $155,240. (11) In May 2015, the United States Bureau of Labor Statistics (BLS) reported that the median annual salary of aircraft engineers was $107,830. (12) Conversely, (13) employment opportunities for aircraft engineers are projected to decrease by 2 percent by 2024. This decrease may be the result of a decline in the manufacturing industry. (14) Nevertheless aircraft engineers who know how to utilize modeling and simulation programs, fluid dynamic software, and robotic engineering tools (15) is projected to remain the most employable.

1. Which of the following would be the best choice for this sentence (reproduced below)?

The knowledge of an aircraft engineer is acquired through years of education, and special (1) licenses are required.

a. NO CHANGE
b. licenses will be required
c. licenses may be required
d. licenses should be required

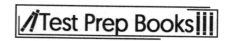

2. Which of the following would be the best choice for this sentence (reproduced below)?

Ideally, an individual will begin his or her preparation for the profession in high school (2) <u>by taking chemistry physics trigonometry and calculus.</u>

a. NO CHANGE
b. by taking chemistry; physics; trigonometry; and calculus.
c. by taking chemistry, physics, trigonometry, and calculus.
d. by taking chemistry, physics, trigonometry, calculus.

3. Which of the following would be the best choice for this sentence (reproduced below)?

Such curricula will aid in (3) <u>one's pursuit of a bachelor's degree</u> in aircraft engineering, which requires several physical and life sciences, mathematics, and design courses.

a. NO CHANGE
b. ones pursuit of a bachelors degree
c. one's pursuit of a bachelors degree
d. ones pursuit of a bachelor's degree

4. Which of the following would be the best choice for this sentence (reproduced below)?

(4) <u>Some of universities provide internship or apprentice opportunities</u> for the students enrolled in aircraft engineer programs.

a. NO CHANGE
b. Some of universities provided internship or apprentice opportunities
c. Some of universities provide internship or apprenticeship opportunities
d. Some universities provide internship or apprenticeship opportunities

5. Which of the following would be the best choice for this sentence (reproduced below)?

(5) <u>These advanced offices oftentimes require a Professional Engineering (PE) license which can be obtained through additional college courses, professional experience, and acceptable scores on the Fundamentals of Engineering (FE) and Professional Engineering (PE) standardized assessments.</u>

a. NO CHANGE
b. These advanced positions oftentimes require acceptable scores on the Fundamentals of Engineering (FE) and Professional Engineering (PE) standardized assessments in order to achieve a Professional Engineering (PE) license. Additional college courses and professional experience help.
c. These advanced offices oftentimes require acceptable scores on the Fundamentals of Engineering (FE) and Professional Engineering (PE) standardized assessments to gain the Professional Engineering (PE) license which can be obtained through additional college courses, professional experience.
d. These advanced positions oftentimes require a Professional Engineering (PE) license which is obtained by acceptable scores on the Fundamentals of Engineering (FE) and Professional Engineering (PE) standardized assessments. Further education and professional experience can help prepare for the assessments.

6. Which of the following would be the best choice for this sentence (reproduced below)?

(6) <u>Once the job begins, this lines of work</u> requires critical thinking, business skills, problem solving, and creativity.

a. NO CHANGE
b. Once the job begins, this line of work
c. Once the job begins, these line of work
d. Once the job begin, this line of work

7. Which of the following would be the best choice for this sentence (reproduced below)?

This level of (7) <u>expertise</u> allows aircraft engineers to apply mathematical equation and scientific processes to aeronautical and aerospace issues or inventions.

a. NO CHANGE
b. expertis
c. expirtise
d. excpertise

8. Which of the following would be the best choice for this sentence (reproduced below)?

This level of expertise (8) <u>allows</u> aircraft engineers to apply mathematical equation and scientific processes to aeronautical and aerospace issues or inventions.

a. NO CHANGE
b. Inhibits
c. Requires
d. Should

9. Which of the following would be the best choice for this sentence (reproduced below)?

This level of expertise allows aircraft engineers to (9) <u>apply mathematical equation and scientific processes</u> to aeronautical and aerospace issues or inventions.

a. NO CHANGE
b. apply mathematical equations and scientific process
c. apply mathematical equation and scientific process
d. apply mathematical equations and scientific processes

10. Which of the following would be the best choice for this sentence (reproduced below)?

(10) <u>For example,</u> aircraft engineers may test, design, and construct flying vessels such as airplanes, space shuttles, and missile weapons.

a. NO CHANGE
b. Therefore,
c. However,
d. Furthermore,

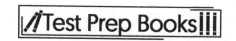

11. Which of the following would be the best choice for this sentence (reproduced below)?

(11) <u>In May 2015, the United States Bureau of Labor Statistics (BLS) reported that the median annual salary of aircraft engineers was $107,830.</u>

a. NO CHANGE
b. May of 2015, the United States Bureau of Labor Statistics (BLS) reported that the median annual salary of aircraft engineers was $107,830.
c. In May of 2015 the United States Bureau of Labor Statistics (BLS) reported that the median annual salary of aircraft engineers was $107,830.
d. In May, 2015, the United States Bureau of Labor Statistics (BLS) reported that the median annual salary of aircraft engineers was $107,830.

12. Which of the following would be the best choice for this sentence (reproduced below)?

(12) <u>Conversely,</u> employment opportunities for aircraft engineers are projected to decrease by 2 percent by 2024.

a. NO CHANGE
b. Similarly,
c. In other words,
d. Accordingly,

13. Which of the following would be the best choice for this sentence (reproduced below)?

Conversely, (13) <u>employment opportunities for aircraft engineers are projected to decrease by 2 percent by 2024.</u>

a. NO CHANGE
b. employment opportunities for aircraft engineers will be projected to decrease by 2 percent by 2024.
c. employment opportunities for aircraft engineers is projected to decrease by 2 percent by 2024.
d. employment opportunities for aircraft engineers was projected to decrease by 2 percent by 2024.

14. Which of the following would be the best choice for this sentence (reproduced below)?

(14) <u>Nevertheless aircraft engineers who know how to utilize</u> modeling and simulation programs, fluid dynamic software, and robotic engineering tools is projected to remain the most employable.

a. NO CHANGE
b. Nevertheless; aircraft engineers who know how to utilize
c. Nevertheless, aircraft engineers who know how to utilize
d. Nevertheless—aircraft engineers who know how to utilize

15. Which of the following would be the best choice for this sentence (reproduced below)?

Nevertheless aircraft engineers who know how to utilize modeling and simulation programs, fluid dynamic software, and robotic engineering tools (15) <u>is projected to remain</u> the most employable.

a. NO CHANGE
b. am projected to remain
c. was projected to remain
d. are projected to remain

Questions 16–30 are based on the following passage:

On September 11th, 2001, a group of terrorists hijacked four American airplanes. The terrorists crashed the planes into the World Trade Center in New York City, the Pentagon in Washington D.C., and a field in Pennsylvania. Nearly 3,000 people died during the attacks, which propelled the United States into a (16) <u>"War on Terror"</u>.

About the Terrorists

(17) <u>Terrorists commonly uses fear and violence to achieve political goals</u>. The nineteen terrorists who orchestrated and implemented the attacks of September 11th were militants associated with al-Qaeda, an Islamic extremist group founded by Osama bin Laden, Abdullah Azzam, and others in the late 1980s. (18) <u>Bin Laden orchestrated the attacks as a response to what he felt was American injustice against Islam and hatred towards Muslims.</u> In his words, "Terrorism against America deserves to be praised."

Islam is the religion of Muslims, (19) <u>who live mainly in south and southwest Asia</u> and Sub-Saharan Africa. The majority of Muslims practice Islam peacefully. However, fractures in Islam have led to the growth of Islamic extremists who strictly oppose Western influences. They seek to institute stringent Islamic law and destroy those who violate Islamic code.

In November 2002, bin Laden provided the explicit motives for the 9/11 terror attacks. According to this list, (20) <u>Americas support of Israel,</u> military presence in Saudi Arabia, and other anti-Muslim actions were the causes.

The Timeline of the Attacks

The morning of September 11 began like any other for most Americans. Then, at 8:45 a.m., a Boeing 767 plane (21) <u>crashed into the north tower of the World Trade Center</u> in New York City. Hundreds were instantly killed. Others were trapped on higher floors. The crash was initially thought to be a freak accident. When a second plane flew directly into the south tower eighteen minutes later, it was determined that America was under attack.

At 9:45 a.m., (22) <u>slamming into the Pentagon was a third plane,</u> America's military headquarters in Washington D.C. The jet fuel of this plane caused a major fire and partial building collapse that resulted in nearly 200 deaths. By 10:00 a.m., the south tower of the World Trade Center collapsed. Thirty minutes later, the north tower followed suit.

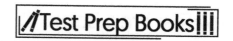

While this was happening, a fourth plane that departed from New Jersey, United Flight 93, was hijacked. The passengers learned of the attacks that occurred in New York and Washington D.C. and realized that they faced the same fate as the other planes that crashed. The passengers were determined to overpower the terrorists in an effort to prevent the deaths of additional innocent American citizens. Although the passengers were successful in (23) <u>diverging</u> the plane, it crashed in a western Pennsylvania field and killed everyone on board. The plane's final target remains uncertain, (24) <u>but believed by many people was the fact that United Flight 93 was heading for the White House.</u>

Heroes and Rescuers

(25) <u>Close to 3,000 people died in the World Trade Center attacks.</u> This figure includes 343 New York City firefighters and paramedics, 23 New York City police officers, and 37 Port Authority officers. Nevertheless, thousands of men and women in service worked (26) <u>valiantly</u> to evacuate the buildings, save trapped workers, extinguish infernos, uncover victims trapped in fallen rubble, and tend to nearly 10,000 injured individuals.

About 300 rescue dogs played a major role in the after-attack salvages. Working twelve-hour shifts, the dogs scoured the rubble and alerted paramedics when they found signs of life. While doing so, the dogs served as a source of comfort and therapy for the rescue teams.

Initial Impacts on America

The attacks of September 11, 2001 resulted in the immediate suspension of all air travel. No flights could take off from or land on American soil. (27) <u>American airports and airspace closed to all national and international flights.</u> Therefore, over five hundred flights had to turn back or be redirected to other countries. Canada alone received 226 flights and thousands of stranded passengers. (28) <u>Needless to say, as cancelled flights are rescheduled, air travel became backed up and chaotic for quite some time.</u>

At the time of the attacks, George W. Bush was the president of the United States. President Bush announced that "We will make no distinction between the terrorists who committed these acts and those who harbor them." The rate of hate crimes against American Muslims spiked, despite President Bush's call for the country to treat them with respect.

Additionally, relief funds were quickly arranged. The funds were used to support families of the victims, orphaned children, and those with major injuries. In this way, the tragic event brought the citizens together through acts of service towards those directly impacted by the attack.

Long-Term Effects of the Attacks

Over the past fifteen years, the attacks of September 11th have transformed the United States' government, travel safety protocols, and international relations. Anti-terrorism legislation became a priority for many countries as law enforcement and intelligence agencies teamed up to find and defeat alleged terrorists.

Present George W. Bush announced a War on Terror. He (29) <u>desired</u> to bring bin Laden and al-Qaeda to justice and prevent future terrorist networks from gaining strength. The War in Afghanistan began in October of 2001 when the United States and British forces bombed al-Qaeda camps. (30) <u>The Taliban, a group of fundamental Muslims who protected Osama bin Laden, was overthrown on December 9, 2001. However, the war continued in order to defeat insurgency campaigns in neighboring countries.</u> Ten years later, the United State Navy SEALS killed Osama bin Laden in Pakistan. During 2014, the United States declared the end of its involvement in the War on Terror in Afghanistan.

Museums and memorials have since been erected to honor and remember the thousands of people who died during the September 11th attacks, including the brave rescue workers who gave their lives in the effort to help others.

16. Which of the following would be the best choice for this sentence (reproduced below)?

Nearly 3,000 people died during the attacks, which propelled the United States into a (16) <u>"War on Terror"</u>.

a. NO CHANGE
b. "war on terror".
c. "war on terror."
d. "War on Terror."

17. Which of the following would be the best choice for this sentence (reproduced below)?

(17) <u>Terrorists commonly uses fear and violence to achieve political goals.</u>

a. NO CHANGE
b. Terrorist's commonly use fear and violence to achieve political goals.
c. Terrorists commonly use fear and violence to achieve political goals.
d. Terrorists commonly use fear and violence to achieves political goals.

18. Which of the following would be the best choice for this sentence (reproduced below)?

(18) <u>Bin Laden orchestrated the attacks as a response to what he felt was American injustice against Islam and hatred towards Muslims.</u>

a. NO CHANGE
b. Bin Laden orchestrated the attacks as a response to what he felt was American injustice against Islam, and hatred towards Muslims.
c. Bin Laden orchestrated the attacks, as a response to what he felt was American injustice against Islam and hatred towards Muslims.
d. Bin Laden orchestrated the attacks as responding to what he felt was American injustice against Islam and hatred towards Muslims.

19. Which of the following would be the best choice for this sentence (reproduced below)?

Islam is the religion of Muslims, (19) <u>who live mainly in south and southwest Asia</u> and Sub-Saharan Africa.

a. NO CHANGE
b. who live mainly in the South and Southwest Asia
c. who live mainly in the south and Southwest Asia
d. who live mainly in the south and southwest asia

20. Which of the following would be the best choice for this sentence (reproduced below)?

According to this list, (20) <u>Americas support of Israel,</u> military presence in Saudi Arabia, and other anti-Muslim actions were the causes.

a. NO CHANGE
b. America's support of israel,
c. Americas support of Israel
d. America's support of Israel,

21. Which of the following would be the best choice for this sentence (reproduced below)?

Then, at 8:45 a.m., a Boeing 767 plane (21) <u>crashed into the north tower of the World Trade Center</u> in New York City.

a. NO CHANGE
b. crashes into the north tower of the World Trade Center
c. crashing into the north tower of the World Trade Center
d. crash into the north tower of the World Trade Center

22. Which of the following would be the best choice for this sentence (reproduced below)?

At 9:45 a.m., (22) <u>slamming into the Pentagon was a third plane,</u> America's military headquarters in Washington D.C.

a. NO CHANGE
b. into the Pentagon slammed a third plane,
c. a third plane slammed into the Pentagon,
d. the Pentagon was slamming by a third plane,

23. Which of the following would be the best choice for this sentence (reproduced below)?

Although the passengers were successful in (23) <u>diverging</u> the plane, it crashed in a western Pennsylvania field and killed everyone on board.

a. NO CHANGE
b. Diverting
c. Converging
d. Distracting

191

24. Which of the following would be the best choice for this sentence (reproduced below)?

The plane's final target remains uncertain, (24) but believed by many people was the fact that United Flight 93 was heading for the White House.

a. NO CHANGE
b. but many believe that United Flight 93 was heading for the White House.
c. also heading for the white house United Flight 93 was believed to be.
d. then many believe that United Flight 93 was heading for the White House.

25. Which of the following would be the best choice for this sentence (reproduced below)?

(25) Close to 3,000 people died in the World Trade Center attacks.

a. NO CHANGE
b. 3,000 people in the World Trade Center attacks died.
c. Dying in the World Trade Center attacks were around 3,000 people.
d. In the World Trade Center attacks were around 3,000 people dying.

26. Which of the following would be the best choice for this sentence (reproduced below)?

Nevertheless, thousands of men and women in service worked (26) valiantly to evacuate the buildings, save trapped workers, extinguish infernos, uncover victims trapped in fallen rubble, and tend to nearly 10,000 injured individuals.

a. NO CHANGE
b. valiently
c. valently
d. vanlyantly

27. Which of the following would be the best choice for this sentence (reproduced below)?

(27) American airports and airspace closed to all national and international flights.

a. NO CHANGE
b. American airports and airspace close to all national and international flights.
c. American airports and airspaces closed to all national and international flights.
d. American airspace and airports were closed to all national and international flights.

28. Which of the following would be the best choice for this sentence (reproduced below)?

(28) Needless to say, as cancelled flights are rescheduled, air travel became backed up and chaotic for quite some time.

a. NO CHANGE
b. As cancelled flights are rescheduled, air travel became backed up and chaotic for quite some time.
c. Needless to say, as cancelled flights were rescheduled, air travel became backed up and chaotic for quite some time.
d. Needless to say, as cancelled flights are rescheduled, air travel became backed up and chaotic over a period of time.

29. Which of the following would be the best choice for this sentence (reproduced below)?

He (29) <u>desired</u> to bring bin Laden and al-Qaeda to justice and prevent future terrorist networks from gaining strength.

a. NO CHANGE
b. Perceived
c. Intended
d. Assimilated

30. Which of the following would be the best choice for this sentence (reproduced below)?

(30) <u>The Taliban, a group of fundamental Muslims who protected Osama bin Laden, was overthrown on December 9, 2001. However, the war continued in order to defeat insurgency campaigns in neighboring countries.</u>

a. NO CHANGE
b. The Taliban was overthrown on December 9, 2001. They were a group of fundamental Muslims who protected Osama bin Laden. However, the war continued in order to defeat insurgency campaigns in neighboring countries.
c. The Taliban, a group of fundamental Muslims who protected Osama bin Laden, on December 9, 2001 was overthrown. However, the war continued in order to defeat insurgency campaigns in neighboring countries.
d. Osama bin Laden's fundamental Muslims who protected him were called the Taliban and overthrown on December 9, 2001. Yet the war continued in order to defeat the insurgency campaigns in neighboring countries.

Questions 31–46 are based on the following passage:

(31) <u>Seeing a lasting social change for African American people Fred Hampton desired to see</u> through nonviolent means and community recognition. (32) <u>As a result, he became an African American activist</u> during the American Civil Rights Movement and led the Chicago chapter of the Black Panther Party.

Hampton's Education

(33) <u>Born and raised in Maywood of Chicago, Illinois in 1948 was Hampton.</u> (34) <u>He was gifted academically and a natural athlete,</u> he became a stellar baseball player in high school. (35) <u>After graduating from Proviso East High School in 1966, he later went on to study law at Triton Junior College.</u>

While studying at Triton, Hampton joined and became a leader of the National Association for the Advancement of Colored People (NAACP). (36) <u>The NAACP gained more than 500 members resulting from his membership.</u> Hampton worked relentlessly to acquire recreational facilities in the neighborhood and improve the educational resources provided to the impoverished black community of Maywood.

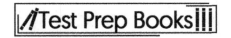

The Black Panthers

The Black Panther Party (BPP) was another activist group that formed around the same time as the NAACP. Hampton was quickly attracted to the Black Panther's approach to the fight for equal rights for African Americans. (37) Hampton eventually joined the chapter, and relocated to downtown Chicago to be closer to its headquarters.

His charismatic personality, organizational abilities, sheer determination, and rhetorical skills enabled him to (38) quickly risen through the chapter's ranks. (39) Hampton soon became the leader of the Chicago chapter of the BPP where he organized rallies, taught political education classes, and established a free medical clinic. He also took part in the community police supervision project and played an instrumental role in the BPP breakfast program for impoverished African American children.

(40) Leading the BPP Hampton's greatest achievement may be his fight against street gang violence in Chicago. In 1969, (41) Hampton was held by a press conference where he made the gangs agree to a nonaggression pact known as the Rainbow Coalition. As a result of the pact, a multiracial alliance between blacks, Puerto Ricans, and poor youth was developed.

Assassination

As the Black Panther Party's popularity and influence grew, the Federal Bureau of Investigation (FBI) placed the group under constant surveillance. In an attempt to (42) neutralize the party, the FBI launched several harassment campaigns against the BPP, raided its headquarters in Chicago three times, and arrested over one hundred of the group's members. (43) During such a raid that occurred Hampton was shot on the morning of December 4th 1969.

In 1976, seven years after the event, it was revealed that William O'Neal, Hampton's trusted bodyguard, was an undercover FBI agent. (44) O'Neal provided the FBI with detailed floor plans of the BPP's headquarters, identifying the exact location of Hampton's bed. (45) It was because of these floor plans that the police were able to target and kill Hampton.

The assassination of Hampton fueled outrage amongst the African American community. It was not until years after the assassination that the police admitted wrongdoing. The Chicago City Council now (46) commemorates December 4th as Fred Hampton Day.

31. Which of the following would be the best choice for this sentence (reproduced below)?

(31) Seeing a lasting social change for African American people Fred Hampton desired to see through nonviolent means and community recognition.

a. NO CHANGE
b. Desiring to see a lasting social change for African American people, Fred Hampton
c. Fred Hampton desired to see lasting social change for African American people
d. Fred Hampton desiring to see last social change for African American people

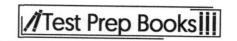

32. Which of the following would be the best choice for this sentence (reproduced below)?

(32) <u>As a result, he became an African American activist</u> during the American Civil Rights Movement and led the Chicago chapter of the Black Panther Party.

a. NO CHANGE
b. As a result he became an African American activist
c. As a result: he became an African American activist
d. As a result of, he became an African American activist

33. Which of the following would be the best choice for this sentence (reproduced below)?

(33) <u>Born and raised in Maywood of Chicago, Illinois in 1948 was Hampton.</u>

a. NO CHANGE
b. Hampton was born and raised in Maywood of Chicago, Illinois in 1948.
c. Hampton is born and raised in Maywood of Chicago, Illinois in 1948.
d. Hampton was born and raised in Maywood of Chicago Illinois in 1948.

34. Which of the following would be the best choice for this sentence (reproduced below)?

(34) <u>He was gifted academically and a natural athlete,</u> he became a stellar baseball player in high school.

a. NO CHANGE
b. A natural athlete and gifted though he was,
c. A natural athlete, and gifted,
d. Gifted academically and a natural athlete,

35. Which of the following would be the best choice for this sentence (reproduced below)?

(35) <u>After graduating from Proviso East High School in 1966, he later went on to study law at Triton Junior College.</u>

a. NO CHANGE
b. He later went on to study law at Triton Junior College graduating from Proviso East High School in 1966.
c. Graduating from Proviso East High School and Triton Junior College went to study.
d. Later at Triton Junior College, studying law, from Proviso East High School in 1966.

36. Which of the following would be the best choice for this sentence (reproduced below)?

(36) <u>The NAACP gained more than 500 members resulting from his membership.</u>

a. NO CHANGE
b. A gain of 500 members happened to the NAACP due to his membership.
c. As a result of his leadership, the NAACP gained more than 500 members.
d. 500 members were gained due to his NAACP membership.

37. Which of the following would be the best choice for this sentence (reproduced below)?

(37) Hampton eventually joined the chapter, and relocated to downtown Chicago to be closer to its headquarters.

a. NO CHANGE
b. Hampton eventually joined the chapter; and relocated to downtown Chicago to be closer to its headquarters.
c. Hampton eventually joined the chapter relocated to downtown Chicago to be closer to its headquarters.
d. Hampton eventually joined the chapter and relocated to downtown Chicago to be closer to its headquarters.

38. Which of the following would be the best choice for this sentence (reproduced below)?

His charismatic personality, organizational abilities, sheer determination, and rhetorical skills enabled him to (38) quickly risen through the chapter's ranks.

a. NO CHANGE
b. quickly rise
c. quickly rose
d. quickly rosed

39. Which of the following would be the best choice for this sentence (reproduced below)?

(39) Hampton soon became the leader of the Chicago chapter of the BPP where he organized rallies, taught political education classes, and established a free medical clinic.

a. NO CHANGE
b. As the leader of the BPP, Hampton: organized rallies, taught political education classes, and established a free medical clinic.
c. As the leader of the BPP, Hampton; organized rallies, taught political education classes, and established a free medical clinic.
d. As the leader of the BPP, Hampton—organized rallies, taught political education classes, and established a medical free clinic.

40. Which of the following would be the best choice for this sentence (reproduced below)?

(40) Leading the BPP Hampton's greatest achievement may be his fight against street gang violence in Chicago.

a. NO CHANGE
b. Greatest achievement of Hampton leading the BPP
c. Hampton's greatest achievement as the leader of the BPP
d. Leader of the BPP Hampton and greatest achievement

41. Which of the following would be the best choice for this sentence (reproduced below)?

In 1969, (41) Hampton was held by a press conference where he made the gangs agree to a nonaggression pact known as the Rainbow Coalition.

a. NO CHANGE
b. a press conference held Hampton
c. held by a press conference was Hampton
d. Hampton held a press conference

42. Which of the following would be the best choice for this sentence (reproduced below)?

In an attempt to (42) neutralize the party, the FBI launched several harassment campaigns against the BPP, raided its headquarters in Chicago three times, and arrested over one hundred of the group's members.

a. NO CHANGE
b. Accommodate
c. Assuage
d. Praise

43. Which of the following would be the best choice for this sentence (reproduced below)?

(43) During such a raid that occurred Hampton was shot on the morning of December 4th 1969.

a. NO CHANGE
b. A raid, occurring because Hampton was shot
c. Hampton was shot during such a raid that occurred
d. Such a raid that occurred, Hampton was shot at

44. Which of the following would be the best choice for this sentence (reproduced below)?

(44) O'Neal provided the FBI with detailed floor plans of the BPP's headquarters, identifying the exact location of Hampton's bed.

a. NO CHANGE
b. O'Neal is providing
c. O'Neal provides
d. O'Neal provided

45. Which of the following would be the best choice for this sentence (reproduced below)?

(45) It was because of these floor plans that the police were able to target and kill Hampton.

a. NO CHANGE
b. These floor plans made it possible for the police to target and kill Hampton.
c. For the police to target and kill Hampton, these floor plans made it possible.
d. These floor plans make it possible for the police to target and kill Hampton.

46. Which of the following would be the best choice for this sentence (reproduced below)?

The Chicago City Council now (46) <u>commemorates</u> December 4th as Fred Hampton Day.

    a. NO CHANGE
    b. disparages
    c. exculpates
    d. brandishes

Questions 47–60 are based on the following passage:

A flood occurs when an area of land that is normally dry becomes submerged with water. Floods have affected Earth since the (47) <u>beginning of time, and are caused</u> by many different factors. Flooding can occur slowly or within seconds and can submerge small regions or extend over vast areas of land. Their impact on society and the environment can be harmful or helpful.

What Causes Flooding?

Floods may be caused by natural phenomenon, induced by the activities of humans and other animals, or the failure of an infrastructure. (48) <u>Areas located near bodies of water are prone to flooding as are low-lying regions.</u>

Global warming is the result of air pollution (49) <u>that prevents the suns radiation</u> from being emitted back into space. Instead, the radiation is trapped in Earth and results in global warming. The warming of the Earth has resulted in climate changes. As a result, floods have been occurring with increasing regularity. Some claim that the increased temperatures on Earth may cause the icebergs to melt. They fear that the melting of icebergs will cause the (50) <u>oceans levels</u> to rise and flood coastal regions.

Most commonly, flooding is caused by excessive rain. The ground is not able to absorb all the water produced by a sudden heavy rainfall or rainfall that occurs over a prolonged period of time. Such rainfall (51) <u>may cause to overflow the water in rivers and other bodies of water.</u> The excess water can cause dams to break. Such events can cause flooding of the surrounding riverbanks or coastal regions.

Flash flooding can occur without warning and without rainfall. Flash floods may be caused by a river being blocked by (52) <u>a glacier; avalanche; landslide; logjam; a beaver's obstruction; construction; or dam.</u> Water builds behind such a blockage. Eventually, the mass and force of the built-up water become so extreme that it causes the obstruction to break. Thus, enormous amounts of water rush out towards the surrounding areas.

Areal or urban flooding occurs because the land has become hardened. The hardening of land may result from urbanization or drought. Either way, the hardened land prevents water from seeping into the ground. Instead, the water resides on top of the land.

(53) <u>Finally, flooding may result after severe hurricanes, tsunamis, then tropical cyclones.</u> Local defenses and infrastructures are no matches for the tidal surges and waves caused by these natural phenomena. Such events are bound to result in the flooding of nearby coastal regions or estuaries.

A Flood's After-Effects

Flooding can result in severe devastation of nearby areas. Flash floods and tsunamis can result in sweeping waters that travel at destructive speeds. Fast-moving water has the power to demolish all obstacles in its path such as homes, trees, bridges, and buildings. Animals, plants, and humans may all lose their lives during a flood.

Floods can also cause pollution and infection. Sewage may seep from drains or septic tanks and contaminate drinking water or surrounding lands. Similarly, toxins, fuels, debris from annihilated buildings, and other hazardous materials can leave water unusable for consumption. (54) <u>As the water begins to drain, mold may begin to grow.</u> As a result, residents of flooded areas may be left without power, drinkable water, or be exposed to toxins and other diseases.

(55) <u>Although often associated with devastation, not all flooding results</u> in adverse circumstances. (56) <u>For thousands of years, people have inhabited floodplains of rivers.</u> (57) <u>Examples include the Mississippi Valley of the United States, the Nile River in Egypt, and the Tigris River of the Middle East</u>. The flooding of such rivers (58) <u>caused</u> nutrient-rich silts to be deposited on the floodplains. Thus, after the floods recede, an extremely fertile soil is left behind. This soil is conducive to the agriculture of bountiful crops and has sustained the diets of humans for millennium.

Proactive Measures Against Flooding

Technologies now allow scientists to predict where and when flooding is likely to occur. Such technologies can also be used (59) <u>to project</u> the severity of an anticipated flood. In this way, local inhabitants can be warned and take preventative measures such as boarding up their homes, gathering necessary provisions, and moving themselves and possessions to higher grounds.

The (60) <u>picturesque</u> views of coastal regions and rivers have long enticed people to build near such locations. Due to the costs associated with the repairs needed after the flooding of such residencies, many governments now require inhabitants of flood-prone areas to purchase flood insurance and build flood-resistant structures. Pictures of all items within a building or home should be taken so that proper reimbursement for losses can be made in the event that a flood does occur.

Staying Safe During a Flood

If a forecasted flood does occur, (60) <u>so people</u> should retreat to higher ground such as a mountain or roof. Flooded waters may be contaminated, contain hidden debris, or travel at high speeds. Therefore, people should not attempt to walk or drive through a flooded area. To prevent electrocution, electrical outlets and downed power lines need to be avoided.

The Flood Dries Up

Regardless of the type or cause of a flood, floods can result in detrimental alterations to nearby lands and serious injuries to nearby inhabitants. By understanding flood cycles, civilizations can learn to take advantage of flood seasons. By taking the proper

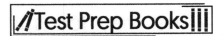

precautionary measures, people can stay safe when floods occur. Thus, proper knowledge can lead to safety and prosperity during such an adverse natural phenomenon.

47. Which of the following would be the best choice for this sentence (reproduced below)?

Floods have affected Earth since the (47) beginning of time, and are caused by many different factors.

a. NO CHANGE
b. beginning of time and are causing
c. beginning of time so to cause
d. beginning of time and are caused

48. Which of the following would be the best choice for this sentence (reproduced below)?

(48) Areas located near bodies of water are prone to flooding as are low-lying regions.

a. NO CHANGE
b. Low-lying regions are prone to flooding, areas located near bodies of water are prone to flooding.
c. Areas located near bodies of water are prone to flooding, as are low-lying regions.
d. Near bodies of water are areas that are prone to flooding and areas near low-lying regions are prone to flooding.

49. Which of the following would be the best choice for this sentence (reproduced below)?

Global warming is the result of air pollution (49) that prevents the suns radiation from being emitted back into space.

a. NO CHANGE
b. that which prevents the sun's radiation
c. the sun's radiation being prevented
d. that prevents the sun's radiation

50. Which of the following would be the best choice for this sentence (reproduced below)?

They fear that the melting of icebergs will cause the (50) oceans levels to rise and flood coastal regions.

a. NO CHANGE
b. ocean levels
c. ocean's levels
d. levels of the oceans

51. Which of the following would be the best choice for this sentence (reproduced below)?

Such rainfall (51) may cause to overflow the water in rivers and other bodies of water.

a. NO CHANGE
b. may been causing the rivers and other bodies of water to overflow.
c. may cause the rivers and other bodies of water to overflow.
d. may cause other bodies of water and water in rivers to overflow.

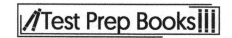

52. Which of the following would be the best choice for this sentence (reproduced below)?

Flash floods may be caused by a river being blocked by (52) <u>a glacier; avalanche; landslide; logjam; a beaver's obstruction; construction; or dam.</u>

a. NO CHANGE
b. a glacier avalanche landslide logjam a beaver's obstruction construction or dam.
c. a glacier, avalanche, landslide, logjam, a beavers obstruction, construction, or dam.
d. a glacier, avalanche, landslide, logjam, a beaver's obstruction, construction, or dam.

53. Which of the following would be the best choice for this sentence (reproduced below)?

(53) <u>Finally, flooding may result after severe hurricanes, tsunamis, then tropical cyclones.</u>

a. NO CHANGE
b. Finally, flooding may result after severe hurricanes, tsunamis, or tropical cyclones.
c. Finally, flooding may result after severe hurricanes, tsunamis, so tropical cyclones.
d. Finally, flooding may result after severe hurricanes, tsunamis, but tropical cyclones.

54. Which of the following would be the best choice for this sentence (reproduced below)?

(54) <u>As the water begins to drain, mold may begin to grow.</u>

a. NO CHANGE
b. As the waters begins to drain, mold may begin to grow.
c. Mold may begin grow as the water begins to drain.
d. The water will begin to drain and mold will begin to grow.

55. Which of the following would be the best choice for this sentence (reproduced below)?

(55) <u>Although often associated with devastation, not all flooding results</u> in adverse circumstances.

a. NO CHANGE
b. Although often associated with devastation not all flooding results
c. Although often associated with devastation. Not all flooding results
d. While often associated with devastation, not all flooding results

56. Which of the following would be the best choice for this sentence (reproduced below)?

(56) <u>For thousands of years, people have inhabited floodplains of rivers.</u>

a. NO CHANGE
b. For thousands of years, people have inhabited river floodplains.
c. For thousands of years people have inhabited river floodplains.
d. For thousand's of years, people have inhabited floodplains of rivers.

57. Which of the following would be the best choice for this sentence (reproduced below)?

(57) <u>Examples include the Mississippi Valley of the United States, the Nile River in Egypt, and the Tigris River of the Middle East.</u>

a. NO CHANGE
b. Examples of floodplains include the Mississippi Valley of the United States, the Nile River in Egypt, and the Tigris River of the Middle East.
c. Examples include the Mississippi Valley of the United States . . . the Nile River in Egypt . . . and the Tigris River of the Middle East.
d. Examples of floodplains include the Mississippi Valley of the United States the Nile River in Egypt and the Tigris River of the Middle East.

58. Which of the following would be the best choice for this sentence (reproduced below)?

The flooding of such rivers (58) <u>caused</u> nutrient-rich silts to be deposited on the floodplains.

a. NO CHANGE
b. Cause
c. Causing
d. Causes

59. Which of the following would be the best choice for this sentence (reproduced below)?

Such technologies can also be used (59) <u>to project</u> the severity of an anticipated flood.

a. NO CHANGE
b. Projecting
c. Project
d. Projected

60. Which of the following would be the best choice for this sentence (reproduced below)?

If a forecasted flood does occur, (60) <u>so people</u> should retreat to higher ground such as a mountain or roof.

a. NO CHANGE
b. then people
c. and people
d. for people

Questions 61–75 are based on the following passage:

In (61) <u>"The Odyssey,"</u> Odysseus develops out of his experiences and the people he meets along his journey home. Many of his encounters involve female characters, some of whom offer Odysseus aid in his journey. (62) <u>However, several of these</u> characters deceive and even pose great danger to the hero. (63) <u>This makes his journey home harder, it forces Odysseus himself to change and adapt</u> in order to deal with the challenges. (64) <u>For the time Odysseus reaches home,</u> he has become notably distrustful of women and even those who have true intentions. It is this sense of caution that

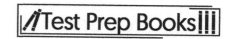

ultimately serves Odysseus in successfully defeating the suitors of Penelope upon his return home.

Odysseus would not have been able to defeat the suitors without stealth and deception. He had (65) <u>to conceal</u> himself in order to achieve revenge. This is something we see earlier in Odysseus' encounter with Polyphemus the Cyclops. While not female, Polyphemus displayed feminine qualities characterized by his "womb-like cave." (66) <u>Entering into the dwelling Odysseus directly demanded hospitality Polyphemus instead butchered his men in spite of custom.</u> In order to survive the encounter, Odysseus (67) <u>relinquishes</u> his true identity by telling Polyphemus his name is "Nobody." After the carnage of his men, he does not entrust the Cyclops with his true name. Rather, Odysseus uses disguise and cunning to trick Polyphemus into reopening the cave. When he emerges, he is then reborn again as "Odysseus."

This pattern is echoed again when Odysseus reaches Ithaca: "I look for endless ground to be spattered by the blood and brains of the suitors, these men who are eating all your substance away. But come now, let me make you so that no mortal can recognize you." Here, Athena reveals her plan to disguise Odysseus as he makes his move against the suitors. Why would Odysseus embrace the idea? With Polyphemus, Odysseus entered the cave trusting he would be received as a welcomed guest, but he wasn't. (68) <u>Clearly, Odysseus isn't making the same mistake twice in trusting people to automatically abide by custom.</u> Using a disguise allows Odysseus to apply strategy in a similar manner he had with Polyphemus. (69) <u>This passage specifically described</u> the suitors as eating away at Odysseus' substance, seeming to further the parallel with Polyphemus who devoured Odysseus' men. (70) <u>Also like with Polyphemus</u>, Odysseus only reveals his true identity when he knows his plan has succeeded. The disguise concept presents a strategic role, but it also sheds further light on the impact of Odysseus' travels. To conceal (71) <u>ones</u> identity is to withhold trust.

The Circe episode matches Odysseus against someone he already knows to be untrustworthy. It is known that Circe welcomes all men who (72) <u>enter upon</u> her island with food and drink, but this is a deception meant to ensnare them. This xenia, or hospitality, that Odysseus would have been accustomed to, turns out to be farce. She violates the trust of her guests by turning them into swine, thus making her a deceitful host and a woman Odysseus cannot trust. In order to (73) <u>assure</u> his and the crew's safety, Odysseus must look past her empty courtesy and deceive her in a way that will remedy the situation. With the knowledge of Circe's dark intentions (and Hermes' instructions), Odysseus attempts to out-maneuver Circe by making her think he will kill her. By doing this, he is taking on deceitful qualities so as to ensure he can bend her to proper behavior, which works. Still untrusting of Circe's submission, Odysseus makes her swear a formal oath: "I would not be willing to go to bed with you unless you can bring yourself, O goddess, to swear me a great oath that there is no other evil hurt you devise against me" (10:342-344). Even though Odysseus tames Circe, he is still distrustful. The oath becomes a final assurance that she is sincere. Until he knows for certain that no more treachery can befall him (the oath), he does not partake in showing any form of trust.

In the Land of the Dead, Odysseus encounters Agamemnon, who describes his own murder at the hand of his wife, Klytaimestra. Not only is this an example of a wife's betrayal, but a betrayal that appears close to Odysseus' own situation. Like Agamemnon, Odysseus (74) is returning home to his wife. However, Agamemnon didn't realize his wife had foul intentions, he trusted her to receive him with open arms:

> See, I had been thinking that I would be welcome to my children and thralls of my household when I came home, but she with thoughts surpassing grisly splashed the shame on herself and on the rest of her sex, on women still to come, even on the one whose acts are virtuous.

Clearly this is a cautionary story for Odysseus. After telling Odysseus of Klytaimestra's betrayal, Agamemnon warns Odysseus that all women are inherently distrustful. By this time, Odysseus has already been deceived and nearly killed by female/female-like characters. Agamemnon's logic seems to back up what he already experienced. (75) As the text progresses, Odysseus encounters the Sirens and Calypso, who seem to corroborate the idea that women are bad news. However, what is most impressionable on Odysseus is Agamemnon's distrust of even virtuous women, "even on the one whose acts are virtuous." Who is to say that they cannot turn against him like Klytaimestra did against Agamemnon. This seems to cement in Odysseus a fear of betrayal.

61. Which of the following would be the best choice for this sentence (reproduced below)?

In (61) "The Odyssey," Odysseus develops out of his experiences and the people he meets along his journey home.

a. NO CHANGE
b. 'The Odyssey'
c. *The Odyssey*
d. The Odyssey

62. Which of the following would be the best choice for this sentence (reproduced below)?

(62) However, several of these characters deceive and even pose great danger to the hero.

a. NO CHANGE
b. However these
c. However several of these
d. Several of these

63. Which of the following would be the best choice for this sentence (reproduced below)?

(63) This makes his journey home harder, it forces Odysseus himself to change and adapt in order to deal with the challenges.

a. NO CHANGE
b. This makes his journey home harder it forces Odysseus himself to change and adapt
c. This makes his journey home harder which forces Odysseus to change and adapt
d. This makes his journey home harder, forcing Odysseus to change and adapt

64. Which of the following would be the best choice for this sentence (reproduced below)?

(64) <u>For the time Odysseus reaches home,</u> he has become notably distrustful of women and even those who have true intentions.

a. NO CHANGE
b. When the time Odysseus reaches home,
c. By the time Odysseus reaches home,
d. At the time Odysseus reaches home,

65. Which of the following would be the best choice for this sentence (reproduced below)?

He had (65) <u>to conceal</u> himself in order to achieve revenge.

a. NO CHANGE
b. To be concealed
c. Conceals
d. Concealing

66. Which of the following would be the best choice for this sentence (reproduced below)?

(66) <u>Entering into the dwelling Odysseus directly demanded hospitality Polyphemus instead butchered his men in spite of custom.</u>

a. NO CHANGE
b. Entering into the dwelling, Odysseus directly demanded hospitality. Polyphemus instead butchered his men in spite of custom.
c. Entering into the dwelling, Odysseus directly demanded hospitality, Polyphemus instead butchered his men in spite of custom.
d. Entering into the dwelling; Odysseus directly demanded hospitality; Polyphemus instead butchered his men in spite of custom.

67. Which of the following would be the best choice for this sentence (reproduced below)?

In order to survive the encounter, Odysseus (67) <u>relinquishes</u> his true identity by telling Polyphemus his name is "Nobody."

a. NO CHANGE
b. Conceals
c. Withholds
d. Surrenders

68. Which of the following would be the best choice for this sentence (reproduced below)?

(68) Clearly, Odysseus isn't making the same mistake twice in trusting people to automatically abide by custom.

a. NO CHANGE
b. Clearly Odysseus isn't making the same mistake twice in trusting people to automatically abide by custom.
c. Clearly, Odysseus isn't making the same mistake twice; trusting people to automatically abide by custom.
d. Odysseus isn't making the same mistake twice in trusting people, clearly, to automatically abide by custom.

69. Which of the following would be the best choice for this sentence (reproduced below)?

(69) This passage specifically described the suitors as eating away at Odysseus' substance, seeming to further the parallel with Polyphemus who devoured Odysseus' men.

a. NO CHANGE
b. This passages specifically describes
c. This passage specifically describes
d. These passage specifically describes

70. Which of the following would be the best choice for this sentence (reproduced below)?

(70) Also like with Polyphemus, Odysseus only reveals his true identity when he knows his plan has succeeded.

a. NO CHANGE
b. As he did with Polyphemus
c. As he did before
d. With the exact method as he had with Polyphemus

71. Which of the following would be the best choice for this sentence (reproduced below)?

To conceal (71) ones identity is to withhold trust.

a. NO CHANGE
b. One's
c. Someone's
d. Oneself

72. Which of the following would be the best choice for this sentence (reproduced below)?

It is known that Circe welcomes all men who (72) enter upon her island with food and drink, but this is a deception meant to ensnare them.

a. NO CHANGE
b. Land upon
c. Arrive on
d. Crash on

25

73. Which of the following would be the best choice for this sentence (reproduced below)?

In order to (73) <u>assure</u> his and the crew's safety, Odysseus must look past her empty courtesy and deceive her in a way that will remedy the situation.

a. NO CHANGE
b. Ensure
c. Prevent
d. Vindicate

74. Which of the following would be the best choice for this sentence (reproduced below)?

Like Agamemnon, Odysseus (74) <u>is returning</u> home to his wife.

a. NO CHANGE
b. returns
c. returned
d. was returned

75. Which of the following would be the best choice for this sentence (reproduced below)?

(75) <u>As the text progresses, Odysseus</u> encounters the Sirens and Calypso, who seem to corroborate the idea that women are bad news.

a. NO CHANGE
b. As the text progresses Odysseus
c. The text progresses with Odysseus
d. As Odysseus progresses in the text

Math

1. At the beginning of the day, Xavier has 20 apples. At lunch, he meets his sister Emma and gives her half of his apples. After lunch, he stops by his neighbor Jim's house and gives him 6 of his apples. He then uses $\frac{3}{4}$ of his remaining apples to make an apple pie for dessert at dinner. At the end of the day, how many apples does Xavier have left?
a. 4
b. 6
c. 2
d. 1
e. 3

2. What is the product of two irrational numbers?
a. Irrational
b. Rational
c. Irrational or rational
d. Complex and imaginary
e. Imaginary

3. The graph shows the position of a car over a 10-second time interval. Which of the following is the correct interpretation of the graph for the interval 1 to 3 seconds?

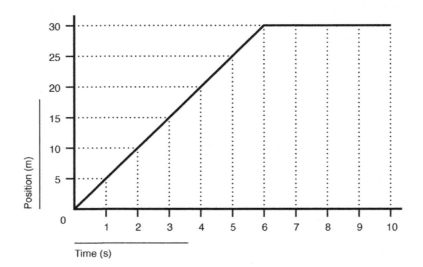

Time (s)

a. The car remains in the same position.
b. The car is traveling at a speed of 5 m/s.
c. The car is traveling up a hill.
d. The car is traveling at 5 mph.
e. The car accelerates at a rate of 5 m/s.

4. Being as specific as possible, how is the number -4 classified?
a. Real, rational, integer, whole, natural
b. Real, rational, integer, natural
c. Real, rational, integer
d. Real, irrational, complex
e. Real, irrational, whole

5. After a 20% discount, Frank purchased a new refrigerator for $850. How much did he save from the original price?
a. $170
b. $212.50
c. $105.75
d. $200
e. $187.50

6. Store brand coffee beans cost $1.23 per pound. A local coffee bean roaster charges $1.98 per 1 ½ pounds. How much more would 5 pounds from the local roaster cost than 5 pounds of the store brand?
a. $0.55
b. $1.55
c. $1.45
d. $0.45
e. $0.35

7. What is the solution to the following problem in decimal form?

$$\frac{3}{5} \times \frac{7}{10} \div \frac{1}{2}$$

    a. 0.042
    b. 84%
    c. 0.84
    d. 0.42
    e. 42%

8. Dwayne has received the following scores on his math tests: 78, 92, 83, and 97. What score must Dwayne get on his next math test to have an overall average of 90?
    a. 89
    b. 98
    c. 95
    d. 94
    e. 100

9. What are all the factors of 12?
    a. 12, 24, 36
    b. 1, 2, 4, 6, 12
    c. 12, 24, 36, 48
    d. 1, 2, 3, 4, 6, 12
    e. 0, 1, 12

10. Which of the following augmented matrices represents the system of equations below?

$$2x - 3y + z = -5$$
$$4x - y - 2z = -7$$
$$-x + 2z = -1$$

a. $\begin{bmatrix} 2 & -3 & 1 & 5 \\ 4 & -1 & 0 & -7 \\ -1 & 0 & 2 & 1 \end{bmatrix}$

b. $\begin{bmatrix} 2 & 4 & -1 \\ -3 & -1 & 0 \\ 1 & -2 & 2 \\ -5 & -7 & -1 \end{bmatrix}$

c. $\begin{bmatrix} 2 & 4 & -1 & -5 \\ -3 & -1 & 0 & -7 \\ 2 & -2 & 2 & -1 \end{bmatrix}$

d. $\begin{bmatrix} 2 & -3 & 1 \\ 4 & -1 & -2 \\ -1 & 0 & 2 \end{bmatrix}$

e. $\begin{bmatrix} 2 & -3 & 1 & -5 \\ 4 & -1 & -2 & -7 \\ -1 & 0 & 2 & -1 \end{bmatrix}$

11. What are the zeros of the function: $f(x) = x^3 + 4x^2 + 4x$?
   a. -2
   b. 0, -2
   c. 2
   d. 0, 2
   e. 0

12. If $g(x) = x^3 - 3x^2 - 2x + 6$ and $f(x) = 2$, then what is $g(f(x))$?
   a. -26
   b. 6
   c. $2x^3 - 6x^2 - 4x + 12$
   d. -2
   e. $2x^2 - 6$

13. What is the solution to the following system of equations?
$$x^2 - 2x + y = 8$$
$$x - y = -2$$
   a. $(-2, 3)$
   b. There is no solution.
   c. $(-2, 0) \ (1, 3)$
   d. $(-2, 0) \ (3, 5)$
   e. $(2, 0) \ (-1, 3)$

14. Which of the following is the result after simplifying the expression: $(7n + 3n^3 + 3) + (8n + 5n^3 + 2n^4)$?
   a. $9n^4 + 15n - 2$
   b. $2n^4 + 5n^3 + 15n - 2$
   c. $9n^4 + 8n^3 + 15n$
   d. $2n^4 + 8n^3 + 15n + 3$
   e. $2n^4 + 5n^3 + 15n - 3$

15. What is the product of the following expression?
$$(4x - 8)(5x^2 + x + 6)$$
   a. $20x^3 - 36x^2 + 16x - 48$
   b. $6x^3 - 41x^2 + 12x + 15$
   c. $20x^4 + 11x^2 - 37x - 12$
   d. $2x^3 - 11x^2 - 32x + 20$
   e. $20x^3 - 40x^2 + 24x - 48$

16. How could the following equation be factored to find the zeros?
$$y = x^3 - 3x^2 - 4x$$
   a. $0 = x^2(x - 4), x = 0, 4$
   b. $0 = 3x(x + 1)(x + 4), x = 0, -1, -4$
   c. $0 = x(x + 1)(x + 6), x = 0, -1, -6$
   d. $0 = 3x(x + 1)(x - 4), x = 0, 1, -4$
   e. $0 = x(x + 1)(x - 4), x = 0, -1, 4$

17. The hospital has a nurse to patient ratio of 1:25. If there is a maximum of 325 patients admitted at a time, how many nurses are there?

    a. 13 nurses

    b. 25 nurses

    c. 325 nurses

    d. 12 nurses

    e. 5 nurses

18. Which of the following is the solution for the given equation?

$$\frac{x^2 + x - 30}{x - 5} = 11$$

    a. $x = -6$

    b. All real numbers.

    c. $x = 16$

    d. $x = 5$

    e. There is no solution.

19. Mom's car drove 72 miles in 90 minutes. How fast did she drive in feet per second?

    a. 0.8 feet per second

    b. 48.9 feet per second

    c. 0.009 feet per second

    d. 70.4 feet per second

    e. 21.3 feet per second

20. Solve $V = lwh$ for h.

    a. $lwV = h$

    b. $h = \frac{V}{lw}$

    c. $h = \frac{Vl}{w}$

    d. $h = \frac{Vw}{l}$

    e. $h = \frac{Vl}{w}$

21. What is the domain for the function $y = \sqrt{x}$?

    a. All real numbers

    b. $x \geq 0$

    c. $x > 0$

    d. $y \geq 0$

    e. $x < 0$

22. If Sarah reads at an average rate of 21 pages in four nights, how long will it take her to read 140 pages?

    a. 6 nights

    b. 26 nights

    c. 8 nights

    d. 27 nights

    e. 21 nights

23. The phone bill is calculated each month using the equation $c = 50g + 75$. The cost of the phone bill per month is represented by c, and g represents the gigabytes of data used that month. Identify and interpret the slope of this equation.

    a. 75 dollars per day
    b. 75 gigabytes per day
    c. 50 dollars per day
    d. 50 dollars per gigabyte
    e. The slope cannot be determined

24. What is the function that forms an equivalent graph to $y = \cos(x)$?

    a. $y = \tan(x)$
    b. $y = \csc(x)$
    c. $y = \sin\left(x + \frac{\pi}{2}\right)$
    d. $y = \sin\left(x - \frac{\pi}{2}\right)$
    e. $y = \tan\left(x + \frac{\pi}{2}\right)$

25. If $\sqrt{1 + x} = 4$, what is x?

    a. 10
    b. 15
    c. 20
    d. 25
    e. 36

26. What is the inverse of the function $f(x) = 3x - 5$?

    a. $f^{-1}(x) = \frac{x}{3} + 5$
    b. $f^{-1}(x) = \frac{5x}{3}$
    c. $f^{-1}(x) = 3x + 5$
    d. $f^{-1}(x) = \frac{x+5}{3}$
    e. $f^{-1}(x) = \frac{x}{3} - 5$

27. What are the zeros of $f(x) = x^2 + 4$?

    a. $x = -4$
    b. $x = \pm 2i$
    c. $x = \pm 2$
    d. $x = \pm 4i$
    e. $x = 2, 4$

28. Twenty is 40% of what number?

    a. 60
    b. 8
    c. 200
    d. 70
    e. 50

29. What is the simplified form of the expression $1.2 * 10^{12} \div 3.0 * 10^8$?
   a. $0.4 * 10^4$
   b. $4.0 * 10^4$
   c. $4.0 * 10^3$
   d. $3.6 * 10^{20}$
   e. $4.0 * 10^2$

30. You measure the width of your door to be 36 inches. The true width of the door is 35.75 inches. What is the relative error in your measurement?
   a. 0.7%
   b. 0.007%
   c. 0.99%
   d. 0.1%
   e. 7.0%

31. What is the y-intercept for $y = x^2 + 3x - 4$?
   a. $y = 1$
   b. $y = -4$
   c. $y = 3$
   d. $y = 4$
   e. $y = -3$

32. Is the following function even, odd, neither, or both?
$$y = \frac{1}{2}x^4 + 2x^2 - 6$$
   a. Even
   b. Odd
   c. Neither
   d. Both
   e. Even for all negative x-values and odd for all positive x-values

33. Which equation is not a function?
   a. $y = |x|$
   b. $y = \sqrt{x}$
   c. $x = 3$
   d. $y = 4$
   e. $y = 3x$

34. How could the following function be rewritten to identify the zeros?
$$y = 3x^3 + 3x^2 - 18x$$
   a. $y = 3x(x + 3)(x - 2)$
   b. $y = x(x - 2)(x + 3)$
   c. $y = 3x(x - 3)(x + 2)$
   d. $y = (x + 3)(x - 2)$
   e. $y = 3x(x + 3)(x + 2)$

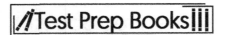

35. A six-sided die is rolled. What is the probability that the roll is 1 or 2?

    a. $\frac{1}{6}$

    b. $\frac{1}{4}$

    c. $\frac{1}{3}$

    d. $\frac{1}{2}$

    e. $\frac{1}{36}$

36. A line passes through the origin and through the point (-3, 4). What is the slope of the line?

    a. $-\frac{4}{3}$

    b. $-\frac{3}{4}$

    c. $\frac{4}{3}$

    d. $\frac{3}{4}$

    e. $\frac{1}{3}$

37. What type of function is modeled by the values in the following table?

x	$f(x)$
1	2
2	4
3	8
4	16
5	32

    a. Linear
    b. Exponential
    c. Quadratic
    d. Cubic
    e. Logarithmic

38. An investment of $2,000 is made into an account with an annual interest rate of 5%, compounded continuously. What is the total value of the investment after eight years?

    a. $4,707
    b. $3,000
    c. $2,983.65
    d. $10,919.63
    e. $1,977.61

39. A ball is drawn at random from a ball pit containing 8 red balls, 7 yellow balls, 6 green balls, and 5 purple balls. What's the probability that the ball drawn is yellow?
   a. $1/26$
   b. $19/26$
   c. $14/26$
   d. 1
   e. $7/26$

40. Two cards are drawn from a shuffled deck of 52 cards. What's the probability that both cards are Kings if the first card isn't replaced after it's drawn?
   a. $1/169$
   b. $1/221$
   c. $1/13$
   d. $4/13$
   e. $1/104$

41. What's the probability of rolling a 6 at least once in two rolls of a die?
   a. $1/3$
   b. $1/36$
   c. $1/6$
   d. $1/12$
   e. $11/36$

42. Given the set $A = \{1, 2, 3, 4, 5, 6, 7, 8, 9, 10\}$ and $B = \{1, 2, 3, 4, 5\}$, what is $A - (A \cap B)$?
   a. $\{6, 7, 8, 9, 10\}$
   b. $\{1, 2, 3, 4, 5\}$
   c. $\{1, 2, 3, 4, 5, 6, 7, 8, 9, 10\}$
   d. $\emptyset$
   e. $\{-1, -2, -3, -4, -5\}$

43. An equilateral triangle has a perimeter of 18 feet. If a square whose sides have the same length as one side of the triangle is built, what will be the area of the square?
   a. 6 square feet
   b. 36 square feet
   c. 256 square feet
   d. 1000 square feet
   e. 324 square feet

44. In a group of 20 men, the median weight is 180 pounds and the range is 30 pounds. If each man gains 10 pounds, which of the following would be true?
    a. The median weight will increase, and the range will remain the same.
    b. The median weight and range will both remain the same.
    c. The median weight will stay the same, and the range will increase.
    d. The median weight and range will both increase.
    e. The median weight will increase, and the range will decrease.

45. For the following similar triangles, what are the values of x and y (rounded to one decimal place)?

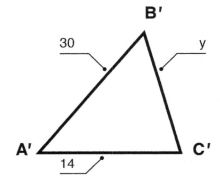

    a. $x = 16.5, y = 25.1$
    b. $x = 19.5, y = 24.1$
    c. $x = 17.1, y = 26.3$
    d. $x = 26.3, y = 17.1$
    e. $x = 24.1, y = 19.5$

46. On Monday, Robert mopped the floor in 4 hours. On Tuesday, he did it in 3 hours. If on Monday, his average rate of mopping was p sq. ft. per hour, what was his average rate on Tuesday?

    a. $\frac{4}{3}p$ sq. ft. per hour

    b. $\frac{3}{4}p$ sq. ft. per hour

    c. $\frac{5}{4}p$ sq. ft. per hour

    d. $p + 1$ sq. ft. per hour

    e. $\frac{1}{3}p$ sq. ft. per hour

47. Which of the following inequalities is equivalent to $3 - \frac{1}{2}x \geq 2$?
   a. $x \geq 2$
   b. $x \leq 2$
   c. $x \geq 1$
   d. $x \leq 1$
   e. $x \leq -2$

48. For which of the following are $x = 4$ and $x = -4$ solutions?
   a. $x^2 + 16 = 0$
   b. $x^2 + 4x - 4 = 0$
   c. $x^2 - 2x - 2 = 0$
   d. $x^2 - x - 16 = 0$
   e. $x^2 - 16 = 0$

49. What are the center and radius of a circle with equation $4x^2 + 4y^2 - 16x - 24y + 51 = 0$?
   a. Center (3, 2) and radius ½
   b. Center (2, 3) and radius ½
   c. Center (3, 2) and radius ¼
   d. Center (2, 3) and radius ¼
   e. Center (2, 2) and radius ¼

50. If the ordered pair $(-3, -4)$ is reflected over the x-axis, what's the new ordered pair?
   a. $(-3, -4)$
   b. $(3, -4)$
   c. $(3, 4)$
   d. $(-3, 4)$
   e. $(-4, -3)$

51. If the volume of a sphere is 288π cubic meters, what are the radius and surface area of the same sphere?
   a. Radius is 6 meters and surface area is 144π square meters
   b. Radius is 36 meters and surface area is 144π square meters
   c. Radius is 6 meters and surface area is 12π square meters
   d. Radius is 36 meters and surface area is 12π square meters
   e. Radius 12 meters and surface area 144π square meters

52. The triangle shown below is a right triangle. What's the value of x?

13 cm

60°

 a. $x = 1.73$
 b. $x = 0.57$
 c. $x = 13$
 d. $x = 14.73$
 e. $x = 22.52$

53. Ten students take a test. Five students get a 50. Four students get a 70. If the average score is 55, what was the last student's score?
 a. 20
 b. 40
 c. 50
 d. 60
 e. 62

54. A sample data set contains the following values: 1, 3, 5, 7. What's the standard deviation of the set?
 a. 2.58
 b. 4
 c. 6.23
 d. 1.1
 e. 0.25

55. A company invests $50,000 in a building where they can produce saws. If the cost of producing one saw is $40, then which function expresses the amount of money the company pays? The variable y is the money paid and x is the number of saws produced.
 a. $y = 50,000x + 40$
 b. $y + 40 = x - 50,000$
 c. $y = 40x - 50,000$
 d. $y = 40x + 50,000$
 e. $y = 4x + 50,000$

56. A pair of dice is thrown, and the sum of the two scores is calculated. What's the expected value of the roll?

    a. 5
    b. 6
    c. 7
    d. 8
    e. 9

57. A line passes through the point (1, 2) and crosses the y-axis at y = 1. Which of the following is an equation for this line?

    a. $y = 2x$
    b. $y = x + 1$
    c. $x + y = 1$
    d. $y = \frac{x}{2} - 2$
    e. $y = x - 1$

58. $x^4 - 16$ can be simplified to which of the following?

    a. $(x^2 - 4)(x^2 + 4)$
    b. $(x^2 + 4)(x^2 + 4)$
    c. $(x^2 - 4)(x^2 - 4)$
    d. $(x^2 - 2)(x^2 + 4)$
    e. $(x^2 - 2)(x^2 + 2)$

59. $(4x^2y^4)^{\frac{3}{2}}$ can be simplified to which of the following?

    a. $8x^3y^6$

    b. $4x^{\frac{5}{2}}y$

    c. $4xy$

    d. $32x^{\frac{7}{2}}y^{\frac{11}{2}}$

    e. x^3y^6

60. A ball is thrown from the top of a high hill, so that the height of the ball as a function of time is $h(t) = -16t^2 + 4t + 6$, in feet. What is the maximum height of the ball in feet?

    a. 6
    b. 6.25
    c. 6.5
    d. 6.75
    e. 6.8

Reading

Fiction

Questions 1–10 are based on the following passage:

"Mademoiselle Eugénie is pretty—I think I remember that to be her name."

"Very pretty, or rather, very beautiful," replied Albert, "but of that style of beauty which I don't appreciate; I am an ungrateful fellow."

"Really," said Monte Cristo, lowering his voice, "you don't appear to me to be very enthusiastic on the subject of this marriage."

"Mademoiselle Danglars is too rich for me," replied Morcerf, "and that frightens me."

"Bah," exclaimed Monte Cristo, "that's a fine reason to give. Are you not rich yourself?"

"My father's income is about 50,000 francs per annum; and he will give me, perhaps, ten or twelve thousand when I marry."

"That, perhaps, might not be considered a large sum, in Paris especially," said the count; "but everything doesn't depend on wealth, and it's a fine thing to have a good name, and to occupy a high station in society. Your name is celebrated, your position magnificent; and then the Comte de Morcerf is a soldier, and it's pleasing to see the integrity of a Bayard united to the poverty of a Duguesclin; disinterestedness is the brightest ray in which a noble sword can shine. As for me, I consider the union with Mademoiselle Danglars a most suitable one; she will enrich you, and you will ennoble her."

Albert shook his head, and looked thoughtful. "There is still something else," said he.

"I confess," observed Monte Cristo, "that I have some difficulty in comprehending your objection to a young lady who is both rich and beautiful."

"Oh," said Morcerf, "this repugnance, if repugnance it may be called, isn't all on my side."

"Whence can it arise, then? for you told me your father desired the marriage."

"It's my mother who dissents; she has a clear and penetrating judgment, and doesn't smile on the proposed union. I cannot account for it, but she seems to entertain some prejudice against the Danglars."

"Ah," said the count, in a somewhat forced tone, "that may be easily explained; the Comtesse de Morcerf, who is aristocracy and refinement itself, doesn't relish the idea of being allied by your marriage with one of ignoble birth; that is natural enough."

Excerpt from the Count of Monte Cristo by Alexandre Dumas

1. The meaning of the word *repugnance* is closest to:
    a. Strong resemblance
    b. Strong dislike
    c. Extreme shyness
    d. Extreme dissimilarity

2. What can be inferred about Albert's family?
    a. Their finances are uncertain.
    b. Albert is the only son in his family.
    c. Their name is more respected than the Danglars'.
    d. Albert's mother and father both agree on their decisions.

3. What is Albert's attitude towards his impending marriage?
    a. Pragmatic
    b. Romantic
    c. Indifferent
    d. Apprehensive

4. What is the best description of the Count's relationship with Albert?
    a. He's like a strict parent, criticizing Albert's choices.
    b. He's like a wise uncle, giving practical advice to Albert.
    c. He's like a close friend, supporting all of Albert's opinions.
    d. He's like a suspicious investigator, asking many probing questions.

5. Which sentence is true of Albert's mother?
    a. She belongs to a noble family.
    b. She often makes poor choices.
    c. She is primarily occupied with money.
    d. She is unconcerned about her son's future.

6. Based on this passage, what is probably NOT true about French society in the 1800s?
    a. Children often received money from their parents.
    b. Marriages were sometimes arranged between families.
    c. The richest people in society were also the most respected.
    d. People were often expected to marry within their same social class.

7. Why is the Count puzzled by Albert's attitude toward his marriage?
    a. He seems reluctant to marry Eugénie, despite her wealth and beauty.
    b. He is marrying against his father's wishes, despite usually following his advice.
    c. He appears excited to marry someone he doesn't love, despite being a hopeless romantic.
    d. He expresses reverence towards Eugénie, despite being from a higher social class than her.

8. The passage is made up mostly of what kind of text?
    a. Narration
    b. Dialogue
    c. Description
    d. Explanation

9. What does the word *ennoble* mean in the middle of the passage?
   a. To create beauty in another person
   b. To endow someone with wealth
   c. To make someone chaste again
   d. To give someone a noble rank or title

10. Why is the count said to have a "forced tone" in the last paragraph?
   a. Because he is in love with Mademoiselle Eugénie and is trying to keep it a secret.
   b. Because he finally agrees with Albert's point of view but still doesn't understand it.
   c. Because he finally understands Albert's point of view but still doesn't agree with it.
   d. Because he is only pretending that Albert is his friend to get information out of him.

Social Science

Questions 11–20 are based upon the following passage:

"MANKIND being originally equals in the order of creation, the equality could only be destroyed by some subsequent circumstance; the distinctions of rich, and poor, may in a great measure be accounted for, and that without having recourse to the harsh ill sounding names of oppression and avarice. Oppression is often the consequence, but seldom or never the means of riches; and though avarice will preserve a man from being necessitously poor, it generally makes him too timorous to be wealthy.

But there is another and greater distinction for which no truly natural or religious reason can be assigned, and that is, the distinction of men into KINGS and SUBJECTS. Male and female are the distinctions of nature, good and bad the distinctions of heaven; but how a race of men came into the world so exalted above the rest, and distinguished like some new species, is worth enquiring into, and whether they are the means of happiness or of misery to mankind.

In the early ages of the world, according to the scripture chronology, there were no kings; the consequence of which was there were no wars; it is the pride of kings which throw mankind into confusion. Holland without a king hath enjoyed more peace for this last century than any of the monarchical governments in Europe. Antiquity favors the same remark; for the quiet and rural lives of the first patriarchs hath a happy something in them, which vanishes away when we come to the history of Jewish royalty.

Government by kings was first introduced into the world by the Heathens, from whom the children of Israel copied the custom. It was the most prosperous invention the Devil ever set on foot for the promotion of idolatry. The Heathens paid divine honors to their deceased kings, and the Christian world hath improved on the plan by doing the same to their living ones. How impious is the title of sacred majesty applied to a worm, who in the midst of his splendor is crumbling into dust!

As the exalting one man so greatly above the rest cannot be justified on the equal rights of nature, so neither can it be defended on the authority of scripture; for the will of the Almighty, as declared by Gideon and the prophet Samuel, expressly disapproves of government by kings. All anti-monarchical parts of scripture have been very smoothly glossed over in monarchical governments, but they undoubtedly merit the attention of countries, which have their governments yet to form. "Render unto Caesar the things which are Caesar's" is the scripture doctrine of courts, yet it is no support of monarchical government, for the Jews at that time were without a king, and in a state of vassalage to the Romans.

Near three thousand years passed away from the Mosaic account of the creation, till the Jews under a national delusion requested a king. Till then their form of government (except in extraordinary cases, where the Almighty interposed) was a kind of republic administered by a judge and the elders of the tribes. Kings they had none, and it was held sinful to acknowledge any being under that title but the Lord of Hosts. And when a man seriously reflects on the idolatrous homage which is paid to the persons of Kings, he need not wonder, that the Almighty ever jealous of his honor, should disapprove of a form of government which so impiously invades the prerogative of heaven.

Excerpt from "Common Sense" by Thomas Paine

11. According to passage, what role does avarice, or greed, play in poverty?
    a. It can make a man very wealthy.
    b. It is the consequence of wealth.
    c. Avarice can prevent a man from being poor, but too fearful to be very wealthy.
    d. Avarice is what drives a person to be very wealthy

12. Of these distinctions, which does the author believe to be beyond natural or religious reason?
    a. Good and bad
    b. Male and female
    c. Human and animal
    d. King and subjects

13. According to the passage, what are the Heathens responsible for?
    a. Government by kings
    b. Quiet and rural lives of patriarchs
    c. Paying divine honors to their living kings
    d. Equal rights of nature

14. Which of the following best states Paine's rationale for the denouncement of monarchy?
    a. It is against the laws of nature.
    b. It is against the equal rights of nature and is denounced in scripture.
    c. Despite scripture, a monarchal government is unlawful.
    d. Neither the law nor scripture denounce monarchy.

15. Based on the passage, what is the best definition of the word *idolatrous*?
    a. Worshipping heroes
    b. Being deceitful
    c. Sinfulness
    d. Engaging in illegal activities

16. What is the essential meaning of lines 41–44?

> And when a man seriously reflects on the idolatrous homage which is paid to the persons of Kings, he need not wonder, that the Almighty ever jealous of his honor, should disapprove of a form of government which so impiously invades the prerogative of heaven.

    a. God would disapprove of the irreverence of a monarchical government.
    b. With careful reflection, men should realize that heaven is not promised.
    c. God will punish those that follow a monarchical government.
    d. Belief in a monarchical government cannot coexist with belief in God.

17. Based on the passage, what is the best definition of the word *timorous* in the first paragraph?
    a. Being full of fear
    b. A characteristic of shyness
    c. Being able to see through someone
    d. Being full of anger and hatred

18. The author's attitude toward the subject can best be described as:
    a. Indifferent and fatigued
    b. Impassioned and critical
    c. Awed and enchanted
    d. Enraged and sulky

19. The main purpose of the fourth paragraph is:
    a. To persuade the audience that heathens were more advanced than Christians.
    b. To explain how monarchs came into existence and how Christians adopted the same system.
    c. To describe the divination of English monarchs and how it is their birthright to be revered.
    d. To counteract the preceding paragraph by giving proof of the damage monarchs can cause.

20. In the last paragraph, what does the author mean by the "Mosaic account of creation"?
    a. The author means that creation is based on individuals composed of cells of two genetically different types.
    b. The author implies that the work of God is like a mosaic painting due to the various formations of creation.
    c. The author means that the kings of the past developed a system in which to maintain accounts of creation.
    d. The author implies that the recorded history of creation is a collection, collage, or pattern taken from various accounts of cultures.

Humanities

Questions 21–30 are based on the following two passages:

Passage 1

Shakespeare and His Plays

People who argue that William Shakespeare is not responsible for the plays attributed to his name are known as anti-Stratfordians (from the name of Shakespeare's birthplace, Stratford-upon-Avon). The most common anti-Stratfordian claim is that William Shakespeare simply was not educated enough or from a high enough social class to have written plays overflowing with references to such a wide range of subjects like history, the classics, religion, and international culture. William Shakespeare was the son of a glove-maker, he only had a basic grade school education, and he never set foot outside of England—so how could he have produced plays of such sophistication and imagination? How could he have written in such detail about historical figures and events, or about different cultures and locations around Europe? According to anti-Stratfordians, the depth of knowledge contained in Shakespeare's plays suggests a well-traveled writer from a wealthy background with a university education, not a countryside writer like Shakespeare. But in fact, there is not much substance to such speculation, and most anti-Stratfordian arguments can be refuted with a little background about Shakespeare's time and upbringing.

First of all, those who doubt Shakespeare's authorship often point to his common birth and brief education as stumbling blocks to his writerly genius. Although it is true that Shakespeare did not come from a noble class, his father was a very *successful* glove-maker and his mother was from a very wealthy land-owning family—so while Shakespeare may have had a country upbringing, he was certainly from a well-off family and would have been educated accordingly. Also, even though he did not attend university, grade school education in Shakespeare's time was actually quite rigorous and exposed students to classic drama through writers like Seneca and Ovid. It is not unreasonable to believe that Shakespeare received a very solid foundation in poetry and literature from his early schooling.

Next, anti-Stratfordians tend to question how Shakespeare could write so extensively about countries and cultures he had never visited before (for instance, several of his most famous works like *Romeo and Juliet* and *The Merchant of Venice* were set in Italy, on the opposite side of Europe!). But again, this criticism does not hold up under scrutiny. For one thing, Shakespeare was living in London, a bustling metropolis of international trade, the most populous city in England, and a political and cultural hub of Europe. In the daily crowds of people, Shakespeare would certainly have been able to meet travelers from other countries and hear firsthand accounts of life in their home country. And, in addition to the influx of information from world travelers, this was also the age of the printing press, a jump in technology that made it possible to print and circulate books much more easily than in the past. This also allowed for a freer flow of information across different countries, allowing people to read about life and ideas from throughout Europe. One needn't travel the continent in order to learn and write about its culture.

Passage 2

The following passage is from The Shakespeare Problem Restated *by G.G. Greenwood*

Now there is very good authority for saying, and I think the truth is so, that at least two of the plays published among the works of Shakespeare are not his at all; that at least three others contain very little, if any, of his writing; and that of the remainder, many contain long passages that are non-Shakespearean. But when we have submitted them all the crucible of criticism we have a magnificent residuum of the purest gold. Here is the true Shakespeare; here is the great magician who, by a wave of his wand, could transmute brass into gold, or make dry bones live and move and have immortal being. Who was this great magician—this mighty dramatist who was "not of an age, but for all time"? Who was the writer of *Venus* and *Lucrece* and the *Sonnets* and *Lear* and *Hamlet*? Was it William Shakespeare of Stratford, the Player? So it is generally believed, and that hypothesis I had accepted in unquestioning faith till my love of the works naturally led me to an examination of the life of the supposed author of them. Then I found that as I read my faith melted away "into thin air." It was not, certainly, that I had (nor have I now) any wish to disbelieve. I was, and I am, altogether willing to accept the Player as the immortal poet if only my reason would allow me to do so. Why not? . . . But the question of authorship is, nevertheless, a most fascinating one. If it be true, as the Rev. Leonard Bacon wrote that "The great world does not care sixpence who wrote *Hamlet*," the great world must, at the same time, be a very small world, and many of us must be content to be outside it. Having given, then, the best attention I was able to give to the question, and more time, I fear, than I ought to have devoted to it, I was brought to the conclusion, as many others have been, that the man who is, truly enough, designated by Messrs. Garnett and Gosse as a "Stratford rustic" is not the true Shakespeare. . .

That Shakespeare the "Stratford rustic and London actor" should have acquired this learning, this culture, and this polish; that *he* should have travelled into foreign lands, studied the life and topography of foreign cities, and the manners and customs of all sorts and conditions of men; that *he* should have written some half-dozen dramas . . . besides qualifying himself as a professional actor; that *he* should have done all this and a good deal more between 1587 and 1592 is a supposition so wild that it can only be entertained by those who are prepared to accept it as a miracle. "And miracles do not happen!"

21. Which sentence contains the author's thesis in the first passage?
    a. People who argue that William Shakespeare is not responsible for the plays attributed to his name are known as anti-Stratfordians.
    b. But in fact, there is not much substance to such speculation, and most anti-Stratfordian arguments can be refuted with a little background about Shakespeare's time and upbringing.
    c. It is not unreasonable to believe that Shakespeare received a very solid foundation in poetry and literature from his early schooling.
    d. Next, anti-Stratfordians tend to question how Shakespeare could write so extensively about countries and cultures he had never visited before.

22. In the first paragraph in Passage 1, "How could he have written in such detail about historical figures and events, or about different cultures and locations around Europe?" is an example of which of the following?
    a. Hyperbole
    b. Onomatopoeia
    c. Rhetorical question
    d. Appeal to authority

23. In Passage 1, how does the author respond to the claim that Shakespeare was not well-educated because he did not attend university?
    a. By insisting upon Shakespeare's natural genius.
    b. By explaining grade school curriculum in Shakespeare's time.
    c. By comparing Shakespeare with other uneducated writers of his time.
    d. By pointing out that Shakespeare's wealthy parents probably paid for private tutors.

24. In Passage 1, the word *bustling* in the third paragraph most nearly means which of the following?
    a. Busy
    b. Foreign
    c. Expensive
    d. Undeveloped

25. In passage 2, the following sentence is an example of what?

    Here is the true Shakespeare; here is the great magician who, by a wave of his wand, could transmute brass into gold, or make dry bones live and move and have immortal being.

    a. Personification
    b. Metaphor
    c. Simile
    d. Allusion

26. In passage 2, the author's attitude toward Stratfordians can be described as which of the following?
    a. Accepting and forgiving
    b. Uncaring and neutral
    c. Uplifting and admiring
    d. Disbelieving and critical

27. What is the relationship between these two sentences from Passage 2?

    Sentence 1: So it is generally believed, and that hypothesis I had accepted in unquestioning faith till my love of the works naturally led me to an examination of the life of the supposed author of them.

    Sentence 2: Then I found that as I read my faith melted away "into thin air."

    a. Sentence 2 explains the main idea in Sentence 1.
    b. Sentence 2 continues the definition begun in Sentence 1.
    c. Sentence 2 analyzes the comment in Sentence 1.
    d. Sentence 2 is a contrast to the idea in Sentence 1.

28. The writing style of Passage 1 could be best described as what?
    a. Expository
    b. Persuasive
    c. Narrative
    d. Descriptive

29. In passage 2, the word *topography* in the second paragraph most nearly means which of the following?
    a. Climate features of an area.
    b. Agriculture specific to place.
    c. Shape and features of the Earth.
    d. Aspects of humans within society.

30. The authors of the passages differ in their opinion of Shakespeare in that the author of Passage 2
    a. Believes that Shakespeare the actor did not write the plays.
    b. Believes that Shakespeare the playwright did not in act in the plays.
    c. Believes that Shakespeare was both the actor and the playwright.
    d. Believes that Shakespeare was neither the actor nor the playwright.

Natural Science

Questions 31–40 are based on the following passage:

In the quest to understand existence, modern philosophers must question if humans can fully comprehend the world. Classical western approaches to philosophy tend to hold that one can understand something, be it an event or object, by standing outside of the phenomena and observing it. It is then by unbiased observation that one can grasp the details of the world. This seems to hold true for many things. Scientists conduct experiments and record their findings, and thus many natural phenomena become comprehendible. However, several of these observations were possible because humans used tools in order to make these discoveries.

This may seem like an extraneous matter. After all, people invented things like microscopes and telescopes in order to enhance their capacity to view cells or the movement of stars. While humans are still capable of seeing things, the question remains if human beings have the capacity to fully observe and see the world in order to understand it. It would not be an impossible stretch to argue that what humans see through a microscope is not the exact thing itself, but a human interpretation of it.

This would seem to be the case in the "Business of the Holes" experiment conducted by Richard Feynman. To study the way electrons behave, Feynman set up a barrier with two holes and a plate. The plate was there to indicate how many times the electrons would pass through the hole(s). Rather than casually observe the electrons acting under normal circumstances, Feynman discovered that electrons behave in two totally different ways depending on whether or not they are observed. The electrons that were observed had passed through either one of the holes or were caught on the plate as particles. However, electrons that weren't observed acted as waves instead of particles and passed through both holes. This indicated that electrons have a dual nature. Electrons seen by the human eye act like particles, while unseen electrons act like waves of energy.

This dual nature of the electrons presents a conundrum. While humans now have a better understanding of electrons, the fact remains that people cannot entirely perceive how electrons behave without the use of instruments. We can only observe one of the mentioned behaviors, which only provides a partial understanding of the entire function of electrons. Therefore, we're forced to ask ourselves whether the world we observe is objective or if it is subjectively perceived by humans. Or, an alternative question: can man understand the world only through machines that will allow them to observe natural phenomena?

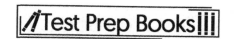

Both questions humble man's capacity to grasp the world. However, those ideas don't consider that many phenomena have been proven by human beings without the use of machines, such as the discovery of gravity. Like all philosophical questions, whether man's reason and observation alone can understand the universe can be approached from many angles.

31. The word *extraneous* in paragraph 2 can be best interpreted as referring to which one of the following?
    a. Indispensable
    b. Bewildering
    c. Superfluous
    d. Exuberant

32. What is the author's motivation for writing the passage?
    a. To bring to light an alternative view on human perception by examining the role of technology in human understanding.
    b. To educate the reader on the latest astroparticle physics discovery and offer terms that may be unfamiliar to the reader.
    c. To argue that humans are totally blind to the realities of the world by presenting an experiment that proves that electrons are not what they seem on the surface.
    d. To reflect on opposing views of human understanding.

33. Which of the following most closely resembles the way in which paragraph four is structured?
    a. It offers one solution, questions the solution, and then ends with an alternative solution.
    b. It presents an inquiry, explains the details of that inquiry, and then offers a solution.
    c. It presents a problem, explains the details of that problem, and then ends with more inquiry.
    d. It gives a definition, offers an explanation, and then ends with an inquiry.

34. For the classical approach to understanding to hold true, which of the following must be required?
    a. A telescope
    b. A recording device
    c. Multiple witnesses present
    d. The person observing must be unbiased

35. Which best describes how the electrons in the experiment behaved like waves?
    a. The electrons moved up and down like actual waves.
    b. The electrons passed through both holes and then onto the plate.
    c. The electrons converted to photons upon touching the plate.
    d. Electrons were seen passing through one hole or the other.

36. The author mentions "gravity" in the last paragraph in order to do what?
    a. To show that different natural phenomena test man's ability to grasp the world.
    b. To prove that since man has not measured it with the use of tools or machines, humans cannot know the true nature of gravity.
    c. To demonstrate an example of natural phenomena humans discovered and understood without the use of tools or machines.
    d. To show an alternative solution to the nature of electrons that humans have not thought of yet.

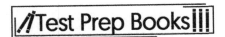
37. Which situation best parallels the revelation of the dual nature of electrons discovered in Feynman's experiment?

a. A man is born color-blind and grows up observing everything in lighter or darker shades. With the invention of special goggles he puts on, he discovers that there are other colors in addition to different shades.

b. The coelacanth was thought to be extinct, but a live specimen was just recently discovered. There are now two living species of coelacanth known to man, and both are believed to be endangered.

c. In the Middle Ages, blacksmiths added carbon to iron, thus inventing steel. The consequences of this important discovery would have its biggest effects during the industrial revolution.

d. In order to better examine and treat broken bones, the x-ray machine was invented and put to use in hospitals and medical centers.

38. Which statement about technology would the author likely disagree with?

a. Technology can help expand the field of human vision.

b. Technology renders human observation irrelevant.

c. Developing tools used in observation and research indicates growing understanding of our world itself.

d. Studying certain phenomena necessitates the use of tools and machines.

39. As it is used in paragraph 4, the word *conundrum* most nearly means:

a. Platitude

b. Enigma

c. Solution

d. Hypothesis

40. What is the author's purpose in paragraph 3?

a. To prove to the audience the thesis of the passage by providing evidence suggesting that electrons behave differently when observed by the human eye.

b. To propose that the experiment conducted was not ethically done and to provide evidence that a new experiment should be conducted in order to reach the truth.

c. To introduce the topic to the audience in a manner that puts it into a practical as well as historical understanding.

d. To pose a question relating to the topic about whether humans fully observe phenomena in an objective or subjective sense.

Science

Passage 1

Questions 1–5 pertain to the following information:

Worldwide, fungal infections of the lung account for significant mortality in individuals with compromised immune function. Three of the most common infecting agents are *Aspergillus, Histoplasma*, and *Candida*. Successful treatment of infections caused by these agents depends on an early and accurate diagnosis. Three tests used to identify specific markers for these mold species include ELISA (enzyme-linked immunosorbent assay), GM Assay (Galactomannan Assay), and PCR (polymerase chain reaction).

Two important characteristics of these tests include sensitivity and specificity. Sensitivity relates to the probability that the test will identify the presence of the infecting agent, resulting in a true positive result. Higher sensitivity equals fewer false-positive results. Specificity relates to the probability that if the test doesn't detect the infecting agent, the test is truly negative for that agent. Higher specificity equals fewer false-negatives.

Figure 1 shows the timeline for the process of infection from exposure to the pathogen to recovery or death.

Figure 1:

Natural History of the Process of Infection

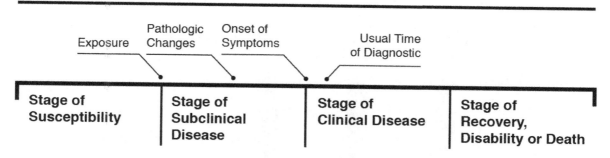

Figure 2 (below) shows the sensitivity and specificity for ELISA, GM assay and PCR related to the diagnosis of infection by *Aspergillus*, *Histoplasma* and *Candida*.

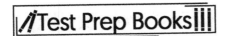

The table below identifies the process of infection in days from exposure for each of the species.

Process of Infection – Days Since Pathogen Exposure			
	Aspergillus	Histoplasma	Candida
Sub-clinical Disease	Day 90	Day 28	Day 7
Detection Possible	Day 118	Day 90	Day 45
Symptoms Appear	Day 145	Day100	Day 120

Figure 3 (below) identifies the point at which each test can detect the organism. Time is measured in days from the time an individual is exposed to the pathogen.

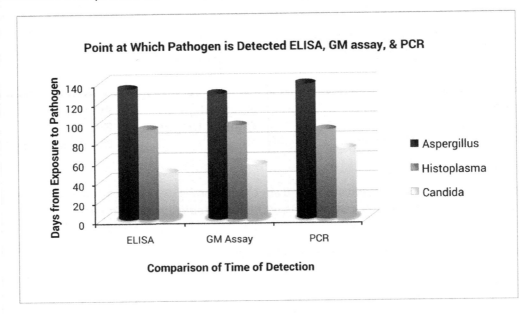

1. Which of the following statements is supported by Figure 2?
   a. For *Candida*, the GM assay will provide the most reliable results.
   b. ELISA testing for *Aspergillus* is the most specific of the three tests.
   c. PCR is the most sensitive method for testing *Histoplasma*.
   d. True positive rates were greater than 75% for all three testing methods.

2. In reference to the table and Figure 3, which pathogen can be detected earlier in the disease process, and by which method?
   a. *Candida* by PCR testing
   b. *Aspergillus* by ELISA testing
   c. *Candida* by GM assay
   d. *Histoplasma* by PCR testing

Test Prep Books!!!

ACT Practice Test #3

3. In reference to Figure 2, which statement is correct?
a. There is a 20% probability that ELISA testing will NOT correctly identify the presence of *Histoplasma*.
b. When GM assay testing for *Candida* is conducted, there is a 31% probability that it will NOT be identified if the organism is present.
c. The probability that GM assay testing for *Aspergillus* will correctly identify the presence of the organism is 99%.
d. The false-negative probabilities for each of the three testing methods identified in Figure 2 indicate that the organism will be detected when present less than 70% of the time.

4. Physicians caring for individuals with suspected *Histoplasma* infections order diagnostic testing prior to instituting treatment. PCR testing results will not be available for 10 days. GM assay results can be obtained more quickly. The physicians opt to wait for the PCR testing. Choose the best possible rationale for that decision.
a. The treatment will be the same regardless of the test results.
b. The individual was not exhibiting any disease symptoms.
c. The probability of PCR testing identifying the presence of the organism is greater than the GM assay.
d. The subclinical disease phase for *Histoplasma* is more than 100 days.

5. Referencing the data in Figures 2 and 3, if ELISA testing costs twice as much as PCR testing, why might it still be the best choice to test for *Candida*?
a. ELISA testing detects the presence of *Candida* sooner than PCR testing.
b. ELISA testing has fewer false-positives than PCR testing.
c. There is only a 69% probability that PCR testing will correctly identify the presence of *Candida*.
d. PCR testing is less sensitive than ELISA testing for *Candida*.

Passage 2

Questions 6–12 pertain to the following information:

Scientists disagree about the cause of Bovine Spongiform Encephalopathy (BSE), also known as "mad cow disease." Two scientists discuss different explanations about the cause of the disease.

Scientist 1

Mad cow disease is a condition that results in the deterioration of brain and spinal cord tissue. This deterioration manifests as sponge-like defects or holes that result in irreversible damage to the brain. The cause of this damage is widely accepted to be the result of an infectious type of protein, called a prion. Normal proteins are located in the cell wall of the central nervous system and function to preserve the myelin sheath around the nerves. Prions are capable of turning normal proteins into other prions by a process that is still unclear, thereby causing the proteins to be "refolded" in abnormal and harmful configurations. Unlike viruses and bacteria, the harmful prions possibly don't contain DNA or RNA, based on the observation of infected tissues in the laboratory that remain infected after immersion in formaldehyde or exposure to ultraviolet light. The transformation from normal to abnormal protein structure and function in a given individual is thought to occur as the result of proteins that are genetically weak or abnormally prone to mutation, or through transmission from another host through food, drugs or organ transplants from infected animals. The abnormal prions also don't trigger an immune response. After prions accumulate in large enough numbers, they form damaging conglomerations that result in the sponge-like holes in tissues, which eventually cause the loss of proper brain function and death.

Figure 1 depicts formation of abnormal prions that results from the abnormal (right) folding of amino acids.

Figure 1:
Configurations of Normal and Abnormal Proteins

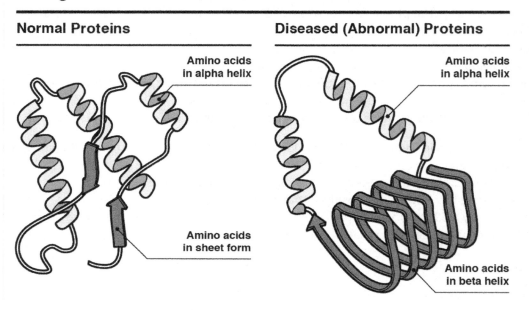

Scientist 2

The degeneration of brain tissue in animals afflicted with mad cow disease is widely considered to be the result of prions. This theory fails to consider other possible causes, such as viruses. Recent studies have shown that infected tissues often contain small particles that match the size and density of viruses. In order to demonstrate that these viral particles are the cause of mad cow disease, researchers used chemicals to inactivate the viruses. When the damaged, inactivated viruses were introduced into healthy tissue, no mad cow disease symptoms were observed. This result indicates that viruses are likely the cause of mad cow disease. In addition, when the infected particles from an infected animal are used to infect a different species, the resulting particles are identical to the original particles. If the infecting agent was a protein, the particles would not be identical because proteins are species-specific. Instead, the infective agent is viewed as some form of a virus that has its own DNA or RNA configuration and can reproduce identical infective particles.

6. Which statement below best characterizes the main difference in the scientists' opinions?
   a. The existence of species-specific proteins
   b. Transmission rates of mad cow disease
   c. The conversion process of normal proteins into prions
   d. The underlying cause of mad cow disease

7. Which of the following statements is INCORRECT?
   a. Scientist 2 proposes that viruses aren't the cause of mad cow disease because chemicals inactivated the viruses.
   b. Scientist 1 suggests that infectious proteins called prions are the cause of mad cow disease.
   c. Scientist 1 indicates that the damaging conglomerations formed by prions eventually result in death.
   d. Scientist 2 reports that infected tissues often contain particles that match the size profile of viruses.

8. which of the following is true according to Scientist 1?
   a. Normal proteins accumulate in large numbers to produce damaging conglomerations.
   b. Prions can change normal proteins into prions.
   c. Species-specific DNA sequences of infected tissues indicate that proteins cause mad cow disease.
   d. Prions are present only in the peripheral nervous system of mammals.

9. Which of the following statements would be consistent with the views of BOTH scientists?
   a. Resulting tissue damage is reversible.
   b. The infecting agent is composed of sheets of amino acids in an alpha helix configuration.
   c. Species-specific DNA can be isolated from infected tissue.
   d. Cross-species transmission of the illness is possible.

10. How does the *conglomeration* described in the passage affect function?
    a. Synapses are delayed
    b. Sponge-like tissue formations occur
    c. Space-occupying lesions compress the nerves
    d. The blood supply to surrounding tissues is decreased

11. What evidence best supports the views of Scientist 2?
    a. Species-specific DNA is present in the infected particles.
    b. Prions are present in the cell membrane.
    c. Prions can trigger an immune response.
    d. The infected particles were inactivated and didn't cause disease.

12. Which of the following statements is supported by this passage?
    a. Scientist 1 favors the claim that viruses are the cause of mad cow disease.
    b. Prions are a type of infectious virus.
    c. The process that results in the formation of the abnormal prion is unclear.
    d. Mad cow disease is caused by normal proteins.

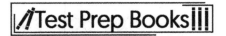

Passage 3

Questions 13–17 pertain to the following information:

Scientists have long been interested in the effect of sleep deprivation on overeating and obesity in humans. Recently, scientists discovered that increased levels of the endocannabinoid 2-Arachidonoylglycerol (2-AG) in the human body is related to overeating. The endocannabinoids play an important role in memory, mood, the reward system, and metabolic processes including glucose metabolism and generation of energy. The endocannabinoid receptors CB1-R and CB2-R are protein receptors located on the cell membrane in the brain, the spinal cord and, to a lesser extent, in the peripheral neurons and the organs of the immune system. The two principal endogenous endocannabinoids are AEA (Anandamide) and 2-Arachidonoylglycerol (2-AG). The endocannabinoids can affect the body's response to chronic stress, mediate the pain response, decrease GI motility, and lessen the inflammatory response in some cancers.

Figure 1 (below) identifies the chemical structure of the endogenous cannabinoids including 2-AG.

Figure 1:
Chemical Structure of Common Endogenous Cannabinoids

The Five-Best known Endocannabinoids Showing the Common 19 - C Backbone Structure and specific R-group Constituents

Recent research has also examined the relationship between sleep deprivation and the levels of 2-AG present in blood, as these conditions relate to obesity. The circadian fluctuations of 2-AG are well-known. Levels normally increase in late afternoon and evening. This physiological increase is thought to contribute to late-day snacking behaviors even after adequate calories have been consumed. The relationship between sleep deprivation and 2-AG appears to relate to the effect of 2-AG on the stress response, represented by sleep deprivation in this study. In order to examine this relationship, university scientists conducted an experiment to identify the influence of injections of 2-AG and sleep deprivation on overeating in a population of non-obese male and female participants that ranged in age from 20–40 years old. To accomplish this, human research subjects (participants) were allowed to eat their favorite junk foods in addition to consuming sufficient calories each day. All of the participants were injected daily with a solution of either sterile normal saline or 2-AG. Daily weight gain was recorded for the three treatment groups that included: participants A–E who received sterile normal saline injections, participants F–J who received 2-AG injections, and participants K–O who received 2-AG injections and were limited to 4.5 hours of sleep each night for 7 nights. The results of the three trials are shown below.

Figure 2 identifies the daily weight gain (in grams) of participants receiving sterile normal saline injections.

Daily Weight Gain for Patients Receiving Sterile Normal Saline Injections

Figure 3 identifies the daily weight gain for participants receiving 2-AG injections.

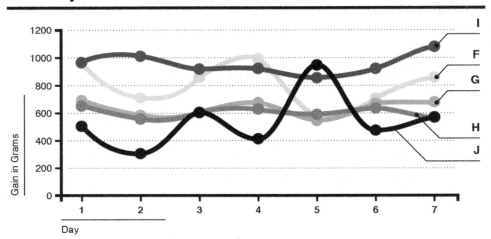

Figure 3:
Daily Weight Gain for Participants Receiving Daily 2-AG Injections

Figure 4 identifies the daily weight gain for participants receiving daily injections of 2-AG who were also limited to 4.5 hours sleep per night for 7 consecutive nights.

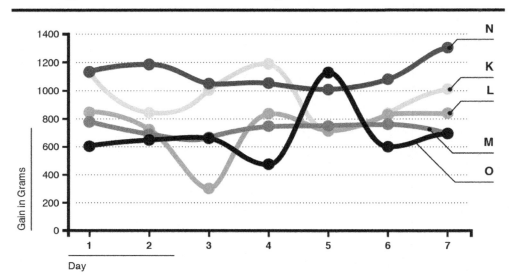

Figure 4:
Daily Weight Gain for Participants Receiving Daily 2-AG Injections Who Were Limited to 4.5 Hours of Sleep Per Night for 7 Consecutive Nights

Figure 5 (below) identifies the participants' average daily weight gain by trial.

13. What was the main hypothesis for this study?
    a. 2-AG injections combined with sleep deprivation will result in weight gain.
    b. 2-AG injections will increase food intake beyond satiety.
    c. Sleep deprivation will result in weight gain.
    d. The placebo effect of the sterile normal saline will influence eating behavior.

14. Do the study results support the hypothesis? Choose the best answer.
    a. No, participants in trials 1 and 3 all gained weight.
    b. Yes, participants in trial 1 gained more weight daily than participants in trial 3.
    c. No, the average weight gain of participants in trial 2 and trial 3 was the same.
    d. Yes, all trial 3 participants gained more weight than trial 1 participants.

15. Describe the study results for participants D and H.
    a. Participant H gained more than one pound each day.
    b. Weight gain for each participant was inconsistent with the study hypothesis.
    c. There was significant fluctuation in the daily weight gain for both participants.
    d. Participant D's average daily weight was two times participant H's average daily weight gain.

16. According to the researchers, which of the following best describes the influence of sleep deprivation on eating behaviors?
    a. The total number of sleep hours is unrelated to the degree of body stress.
    b. Sleep deprivation stimulates the release of endogenous cannabinoids that may increase food intake.
    c. Deprivation of any variety triggers the hunger response.
    d. Sleep deprivation increases eating behaviors in the early morning hours.

17. According to the passage, how does 2-AG influence eating behaviors?
    a. Circadian fluctuations result in increased levels of 2-AG in the afternoon and evening.
    b. Endogenous cannabinoids like 2-AG increase gastric motility, which stimulates the hunger response.
    c. The sedation that results from the presence of 2-AG limits food intake.
    d. Endogenous cannabinoids block the opioid system, which decreases food-seeking behaviors.

Passage 4

Questions 18–22 pertain to the following passage:

A national wholesale nursery commissioned research to conduct a cost/benefit analysis of replacing existing fluorescent grow lighting systems with newer LED lighting systems. LEDs (light-emitting diodes) are composed of various semi-conductor materials that allow the flow of current in one direction. This means that LEDs emit light in a predictable range, unlike conventional lighting systems that give off heat and light in all directions. The wavelength of light of a single LED is determined by the properties of the specific semi-conductor. For instance, the indium gallium nitride system is used for blue, green, and cyan LEDs. As a result, growing systems can be individualized for the specific wavelength requirements for different plant species. In addition, LEDs don't emit significant amounts of heat compared to broadband systems, so plant hydration can be controlled more efficiently.

Figure 1 identifies the visible spectrum with the wavelength expressed in nanometers.

Figure 1:
The Visible Spectrum (Wavelength in Nanometers)

Figure 2 (below) identifies the absorption rates of different wavelengths of light.

Researchers conducted three trials and hypothesized that LEDs would result in greater growth rates than conventional lighting or white light. They also hypothesized that using a combination of red, blue, green, and yellow wavelengths in the LED lighting system would result in a greater growth rate than using red or blue wavelengths alone. Although green and yellow wavelengths are largely reflected by the plant (Figure 2), the absorption rate is sufficient to make a modest contribution to plant growth. Fifteen Impatiens walleriana seed samples were planted in the same growing medium. Temperature, hydration, and light intensity were held constant. Plant height in millimeters was recorded as follows.

Figure 3 identifies the plant growth rate in millimeters with light wavelengths of 440 nanometers.

Figure 3:
Plant Growth Rate (mm) with Light wavelengths of 440 nm

Figure 4 (below) identifies the plant growth rate in mm with light wavelengths of 650 nanometers.

Figure 4:
Plant Growth Rate (mm) with Light wavelengths of 650 nm

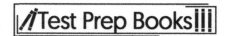

Figure 5 (below) identifies the plant growth rate in millimeters with combined light wavelengths of 440, 550, and 650 nanometers.

Figure 5:
Plant Growth Rate (mm) with Combined Light wavelengths of 440, 550, and 650 nm

Figure 6 (below) identifies average daily plant growth rate in millimeters.

Test Prep Books

ACT Practice Test #3

18. If the minimum plant height required for packaging a plant for sale is 150 millimeters, based on plant growth, how much sooner will the LED plants be packaged compared to the white light plants?
a. 14 days
b. 21 days
c. 35 days
d. 42 days

19. Plants reflect green and yellow light wavelengths. Do the results of the three trials support the view that plants also absorb and use green and yellow light wavelengths for growth?
a. Yes, green and yellow light wavelengths were responsible for plant growth in trial 3.
b. No, white light alone was responsible for measurable plant growth.
c. Yes, the growth rates in trial 3 were greater than the rates in trials 1 and 2.
d. No, only the red and blue wavelengths were effective in stimulating plant growth.

20. When did the greatest rate of growth occur for both groups in trial 1 and trial 2?
a. From 7 days to 14 days
b. From 28 days to 35 days
c. From 21 days to 28 days
d. From 14 days to 21 days

21. If an LED lighting system costs twice as much as a white light system, based only on the average daily growth rate as noted above, would it be a wise investment?
a. No, because multiple different semi-conductors would be necessary.
b. Yes, growth rates are better with LEDs.
c. No, the LED average daily growth rate was not two times greater than the white light rate.
d. Yes, LEDs use less electricity and water.

22. If the researchers conducted an additional trial, trial 4, to measure the effect of green and yellow wavelengths on plant growth, what would be the probable result?
a. The growth rate would exceed trial 1.
b. The growth rate would equal trial 3.
c. The growth rate would be the same as trial 2.
d. The growth rate would be less than trial 1 or trial 2.

Passage 5

Questions 23–28 pertain to the following passage:

Mangoes are a tropical fruit that grow on trees native to Southern Asia called the *Mangifera*. Mangoes are now grown in most frost-free tropical and subtropical locations around the world. India and China harvest the greatest numbers of mangoes. A major problem the mango industry faces each year is the destruction of fruit after harvest. This destruction is the result of spoilage or rotting that occurs during long shipping and storage times.

To prevent the spoilage of mangoes, fruits are stored and shipped in climate-controlled containers. Ideally, mangoes should be stored at around 5 °C, which is about the same temperature as a home refrigerator. Although storage at 5 °C is highly effective at

244

preventing spoilage, the monetary costs associated with maintaining this temperature during long shipping times are prohibitive.

Fruit companies spend large amounts of money to learn about the underlying cause of spoilage and possible methods to prevent loss of their product. Anthracnose, an infection that causes mango decay, is caused by *Colletotrichum,* a type of fungus that has been identified as a major contributor to mango spoilage. This fungus, which may remain dormant on green fruit, grows on the surface of the mango and can penetrate the skin and cause spoilage. The infection first appears during the flowering period as small black dots that progress to dark brown or black areas as ripening occurs. Humidity and excessive rainfall increase the severity of this infection. Previous studies established that colony sizes smaller than 35 millimeters after 4 weeks of travel resulted in acceptable amounts of spoilage.

Currently, several additional pre-treatment measures aimed at prevention are employed to slow decay of the fruit from the harvest to the marketplace. Industry researchers examined the individual and collective benefits of two of these processes, including post-harvest hot water treatment and air cooling at varied transport temperatures in order to identify optimum post-harvest procedures.

Table 1 identifies the observed mango decay in millimeters at 5 °C, 7.5 °C, and 10 °C with two pre-treatment processes over time measured in days since harvest of the fruit.

Table 1: Days Since Harvest

	2	4	6	8	10	12	14	16	18	20	22	24	26	28
5° C														
Water	1	4	7	9	11	12	14	19	22	23	25	27	28	30
Air	0	2	3	6	8	9	11	12	13	15	16	17	18	19
7.5°C														
Water	2	3	4	5	6	8	9	11	12	13	14	15	16	27
Air	0	2	5	6	7	9	10	15	22	23	24	27	32	39
10°C														
Water	2	3	5	7	8	11	12	14	22	35	42	44	47	62
Air	1	2	4	6	7	9	10	15	19	23	27	29	35	44

Figure 1 (below) identifies the observed mango decay of fruit stored at 5 °C measured in millimeters, with two pretreatment processes, over time measured in days since harvest of the fruit.

Figure 1:
5 °C Mango Decay Rates

Figures 2 (below) identifies the observed mango decay of fruit stored at 7.5 °C, measured in millimeters, with two pretreatment processes, over time measured in days since harvest of the fruit.

Figure 2:
7.5 °C Mango Decay Rates

Figures 3 (below) identifies the observed mango decay of fruit stored at 10 °C measured in millimeters, with two pretreatment processes, over time measured in days since harvest of the fruit.

Figures 3:
10 °C Mango Decay Rates

Figure 4 (below) identifies fruit decay measured in millimeters at 5 °C, 7.5 °C, and 10 °C with the combined pre-treatments over 28 days.

Figure 4:
Mango Decay Rates at 5 °C, 7.5 °C, and 10 °C with the Combined Pre-Treatments (Air Cooling & Water Bath)

23. According to the passage above, which of the following statements is false?
    a. The optimal temperature for storing mangos is 5 °C.
    b. *Anacardiaceae Mangifera* is responsible for mango spoilage.
    c. Storing fruit at 5 °C is costly.
    d. Long distance shipping is a critical factor in mango spoilage.

24. If the mangoes were shipped from India to the U.S., and the trip was expected to take 20 days, which model would be best according to the data in Table 1?
    a. The 10 °C model, because fungal levels were acceptable for both pre-treatments.
    b. The 5 °C, model because it's more cost-effective.
    c. The 7.5 °C model, because this temperature is less expensive to maintain, and the fungal levels were acceptable.
    d. No single model is better than the other two models.

25. According to Figures 1–3 above, the largest one-day increase in fruit decay occurred under which conditions?
    a. Air cooling at 10 °C
    b. Hot water wash at 10 °C
    c. Air cooling at 7.5 °C
    d. Hot water wash at 7.5 °C

26. Which pre-treatment method reached unacceptable fungal levels first?
    a. Hot water wash at 7.5 °C
    b. Air cooling at 5 °C
    c. Hot water wash at 10 °C
    d. Air cooling at 10 °C

27. The researchers were attempting to identify the best shipping conditions for mangoes for a 28-day trip from harvest to market. Referencing Figures 1–4, which conditions would be the most cost-effective?
    a. Air cooling pre-treatment at 5 °C
    b. Air cooling and hot water wash pre-treatment at 5 °C
    c. Air cooling pre-treatment at 7.5 °C
    d. Air cooling pre-treatment at 10 °C

28. Shipping mangoes at 5 °C is costly. According to the researchers' findings, is shipping mangoes at 5 °C more cost effective than 7.5 °C for trips lasting more than 28 days when combined air cooling and hot water wash treatments are applied?
    a. Yes, shipping at 7.5 °C combined with both pre-treatments resulted in an unacceptable fungal infection rate.
    b. Yes, fungal infection rates were below 35 mm for both pre-treatments 5 °C.
    c. No, air cooling pre-treatment was acceptable at 10 °C, and it's less expensive to ship fruit at 10 °C.
    d. No, hot water wash rates were lower than air cooling at 5 °C.

Passage 6

Questions 29–34 pertain to the following passage:

Scientists recently discovered that circadian rhythms help regulate sugar consumption by brown adipose tissue. The results of this study suggest that circadian rhythms and fat cells work together to warm the body in preparation for early morning activities involving cold weather. A circadian rhythm refers to life processes controlled by an internal "biological clock" that maintains a 24-hour rhythm. Sleep is controlled by one's circadian rhythm. To initiate sleep, the circadian rhythm stimulates the pineal gland to release the hormone melatonin, which causes sleepiness. Importantly, the circadian rhythm discerns when to begin the process of sleep based on the time of day. During the daytime, sunlight stimulates special cells within the eye, photosensitive retinal ganglion cells, which, in turn, allow the "biological clock" to keep track of how many hours of sunlight there are in a given day.

Brown adipose tissue (BAT) is a type of fat that plays an important role in thermogenesis, a process that generates heat. In humans and other mammals, there are two basic types of thermogenesis: shivering thermogenesis and non-shivering thermogenesis. Shivering thermogenesis involves physical movements, such as shaky hands or clattering teeth. Heat is produced as a result of energy being burned during physical activity. Non-shivering thermogenesis doesn't require physical activity; instead, it utilizes brown adipose tissue to generate heat. Brown fat cells appear dark because they contain large numbers of mitochondria, the organelles that burn sugar to produce energy and heat.

Researchers know that brown adipose tissue (BAT) is essential for maintaining body temperature. A new discovery in humans has shown that circadian rhythms cause BAT to consume more sugar in the early-morning hours. This spike in sugar consumption causes more heat to be produced in BAT. Scientists propose that our human ancestors could have benefited from extra body heat during cold hunts in the morning.

Perhaps more significantly, these new findings may suggest a role for BAT in the prevention of Type 2 Diabetes. Two important questions remain; to what degree does BAT affect blood glucose levels, and is it possible to increase BAT in a given individual? The demonstrated increase in sugar consumption and heat production of BAT is thought to be related to insulin-sensitivity. To examine the first question, researchers conducted three trials to examine the relationship between brown fat and blood glucose levels at different points in the day. PET scanning was used to estimate total body brown fat in 18 non-diabetic participants. Total body brown fat expressed as a proportion of total body fat (either 5%, 10%, or 20%) was the basis for group selection. The researchers hypothesized that the blood glucose levels would be inversely related to the percentage of BAT. Resulting data is included below.

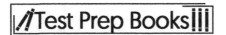

Figure 1 (below) identifies the circadian cycle of blood glucose.

Figure 1:
Normal Circadian Plasma Glucose Levels

Figure 2 (below) identifies the resulting blood glucose measurements for participants with 5% total brown body fat.

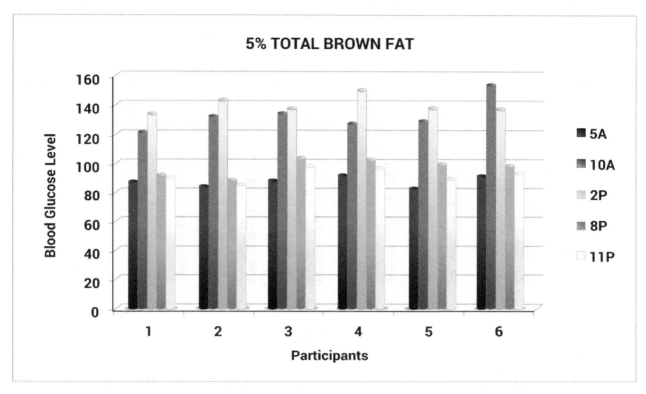

Figure 3 (below) identifies the resulting blood glucose measurements for participants with 10% total brown body fat.

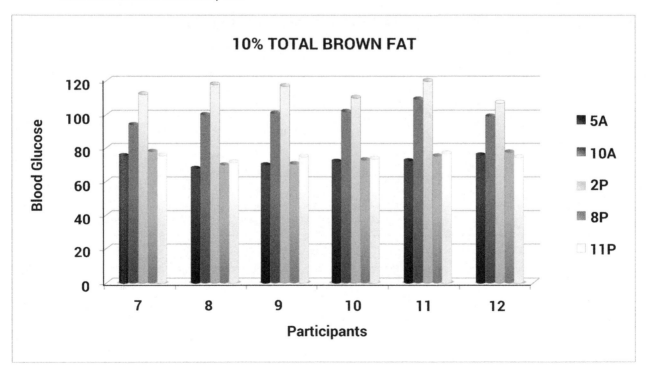

Figure 4 (below) identifies the resulting blood glucose measurements for participants with 20% total brown body fat.

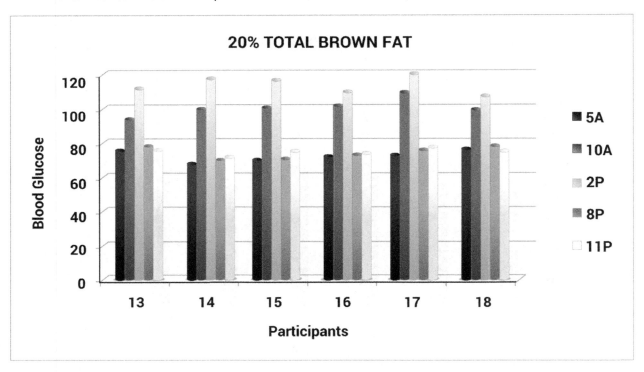

Figure 5 (below) identifies the average blood glucose measurements for the three trials.

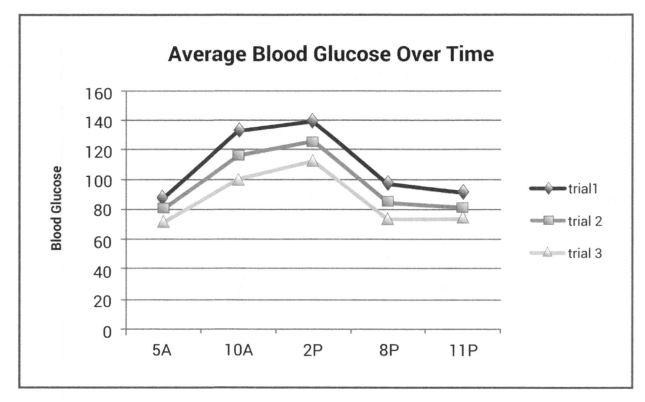

29. Which of the following describes the relationship of the research results in Figure 5?
    a. Positive correlation among the three trials
    b. Curvilinear relationship
    c. Weak negative relationship
    d. No demonstrated relationship

30. Which of the following statements concerning mitochondria is INCORRECT?
    a. Mitochondrial function is diminished in the presence of elevated blood glucose levels.
    b. Mitochondria are responsible for the color of brown fat.
    c. Mitochondria are capable of reproduction in response to energy needs.
    d. Mitochondria are responsible for binding oxygen in mature red blood cells.

31. According to Figure 5, participants' blood sugars were highest at what time of day?
    a. 5 a.m., because heat is generated early in the morning
    b. 8 p.m., because the participants ate less for dinner than lunch
    c. 11 p.m., because brown adipose tissue is not active at night
    d. 2 p.m., because the effects of the early-morning activity of the brown adipose tissue had diminished

32. Circadian rhythms control sleep by doing which of the following?
    a. Stimulating the pineal gland to release ganglia
    b. Stimulating the release of melatonin
    c. Suppressing shivering during cold mornings
    d. Instructing brown adipose tissue to release sugar

33. Which Participant in trial 1 had the highest average blood sugar for the group?
    a. 1
    b. 3
    c. 4
    d. 6

34. Is the data in Figure 5 consistent with the daytime plasma glucose trend in Figure 1?
    a. Yes, Figure 5 blood glucose readings declined from a morning to afternoon.
    b. No, blood glucose readings peaked at 2 p.m..
    c. Yes, morning glucose readings were higher in group 1.
    d. No, nighttime levels fluctuated between 100 and 110.

Passage 7

Questions 35–40 pertain to the following passage:

A biome is a major terrestrial or aquatic environment that supports diverse life forms. Freshwater biomes—including lakes, streams and rivers, and wetlands—account for 0.01% of the Earth's fresh water. Collectively, they are home to 6% of all recognized species. Standing water bodies may vary in size from small ponds to the Great Lakes. Plant life in lakes is specific to the zone of the lake that provides the optimal habitat for a specific species, based on the depth of the water as it relates to light. The photic layer is the shallower layer where light is available for photosynthesis. The aphotic layer is deeper, and the levels of sunlight are too low for photosynthesis. The benthic layer is the bottom-most layer, and its inhabitants are nourished by materials from the photic layer. Light-sensitive cyanobacteria and microscopic algae are two forms of phytoplankton that exist in lakes. As a result of nitrogen and phosphorus from agriculture and sewage run-off, algae residing near the surface can multiply abnormally so that available light is diminished to other species. Oxygen supplies may also be reduced when large numbers of algae die.

Recently, concerns have been raised about the effects of agriculture and commercial development on the quality of national freshwater bodies. In order to estimate the effect of human impact on freshwater, researchers examined plant life from the aphotic layer of three freshwater lakes of approximately the same size located in three different environments. Lake A was located in a remote forested area of western Montana. Lake B was located in central Kansas. Lake C was located in a medium-size city on the west coast of Florida. The researchers hypothesized that the microscopic algae and cyanobacteria populations from Lake A would approach appropriate levels for the size of the lake. They also hypothesized that the remaining two samples would reveal abnormal levels of the phytoplankton. In addition, the researchers measured the concentration of algae at different depths at four different times in another lake identified as having abnormal algae growth. These measurements attempted to identify the point at which light absorption in the photic layer was no longer sufficient for the growth of organisms in the aphotic layer. Resulting data is identified below.

Figure 1 (below) illustrates the zones of the freshwater lake.

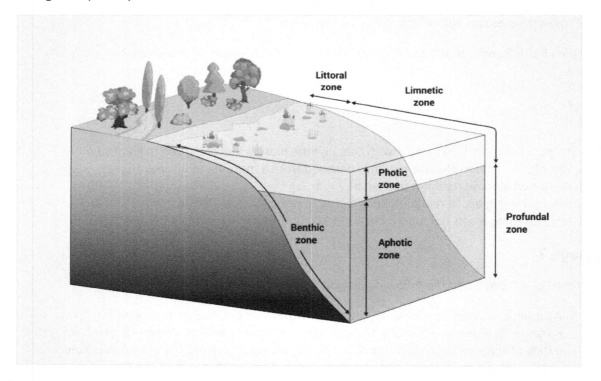

//Test Prep Books|||

Figure 2 (below) identifies algae and cyanobacteria levels in parts per million for Lake A over six measurements.

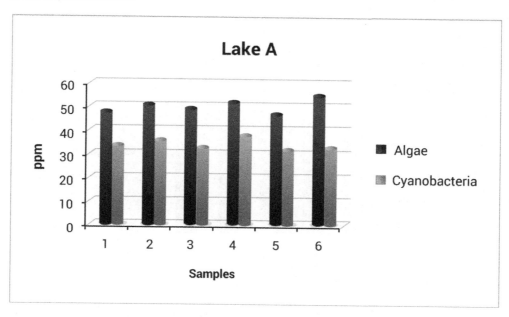

normal: Algae 50 p.p.m. Cyanobacteria 35 p.p.m.

Figure 3 (below) identifies algae and cyanobacteria levels in parts per million for Lake B over six measurements.

normal: Algae 50 p.p.m. Cyanobacteria 35 p.p.m.

Figure 4 (below) identifies algae and cyanobacteria levels in parts per million for Lake C over six measurements.

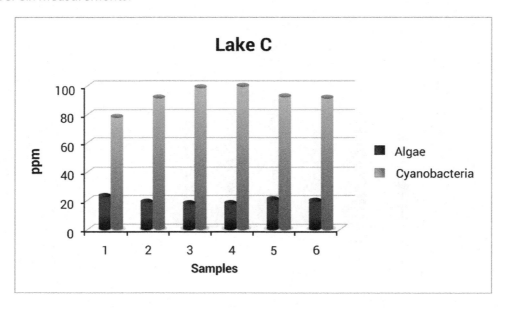

normal: Algae 50 p.p.m. Cyanobacteria 35 p.p.m.

Figure 5 (below) identifies cyanobacteria levels at different depths over time.

35. Based on Figure 2, was the researchers' hypothesis confirmed?
    a. No, the phytoplankton levels were not elevated in the first trial.
    b. Yes, the phytoplankton levels were raised above normal in each sample.
    c. No, the Lake A numbers were normal.
    d. Yes, algae levels were above normal in Lake C.

36. In Lake B, cyanobacteria were decreased and algae were increased. Which of the following is a possible explanation for this finding?
    a. The overgrowth of algae decreased the light energy available for cyanobacteria growth.
    b. Lake B experienced severe flooding, causing the water levels in the lake to rise above normal.
    c. Agricultural chemical residue depleted the food source for cyanobacteria.
    d. Cyanobacteria cannot survive in the cold winter weather in Lake B.

37. What common factor might explain the results for Lake B and Lake C?
    a. Population concentration
    b. Average humidity of the locations
    c. Average heat index
    d. Excess nitrogen and phosphorus in the ground water

38. As algae levels increase above normal, what happens to organisms in the aphotic level?
    a. Growth is limited but sustained.
    b. Species eventually die due to decreased oxygenation.
    c. Cyanobacteria increase to unsafe levels.
    d. Aerobic bacteria multiply.

39. Referencing Figures 3 and 4, which environment would favor organisms in the benthic layer of the corresponding lake?
    a. Figure 4, because the cyanobacteria are protective.
    b. Figure 3, because increased numbers of algae provide more light.
    c. Figure 4, because cyanobacteria are able to survive.
    d. Figure 3, because the levels of both species are normal.

40. Which of the following statements is supported by the data in Figure 5?
    a. Algae growth is greater in July than April.
    b. Cyanobacteria can't exist at 20 meters in this lake.
    c. There's insufficient light in the aphotic layer at 3 meters to support algae growth.
    d. Cyanobacteria growth rates are independent of algae growth at 1 meter.

Writing

Please read the prompt below and answer in an essay format.

Technology has been invading cars for the last several years, but there are some new high tech trends that are pretty amazing. It is now standard in many car models to have a rear-view camera, hands-free phone and text, and a touch screen digital display. Music can be streamed from a paired cell phone, and some displays can even be programmed with a personal photo. Sensors beep to indicate there is something in the driver's path when reversing and changing lanes. Rain-sensing windshield wipers and lights are automatic, leaving the driver with little to do but watch the road and enjoy the ride. The next wave of technology will include cars that automatically parallel park, and a self-driving car is on the horizon. These technological advances make it a good time to be a driver.

1. Analyze and evaluate the passage given.

2. State and develop your own perspective.

3. Explain the relationship between your perspective and the one given.

Answer Explanations for Practice Test #3

English

1. C: The next paragraph states that "These advanced offices oftentimes require a Professional Engineering (PE) license which can be obtained through additional college courses, professional experience, and acceptable scores on the Fundamentals of Engineering (FE) and Professional Engineering (PE) standardized assessments." Since the word *oftentimes* is used instead of *always*, Choice *C* is the best response.

2. C: The best answer is Choice *C*. Items in a list should be separated by a comma. Choice *A* is incorrect because there are no commas within the list to separate the items. Choice *B* is incorrect; a semicolon is used in a series only when a comma is present within the list itself. Choice *D* is incorrect because the conjunction *and* is missing before the word *calculus*.

3. A: The sentence is correct as-is. The words *one* and *bachelor* have apostrophe -*s* at the end because they show possession for the words that come after. The other answer choices do not indicate possession is being shown.

4. D: To begin, *of* is not required here. *Apprenticeship* is also more appropriate in this context than *apprentice opportunities*; *apprentice* describes an individual in an apprenticeship, not an apprenticeship itself. Both of these changes are needed, making Choice *D* the correct answer.

5. D: Choice *D* is correct because it breaks the section into coherent sentences and emphasizes the main point the author is trying to communicate: the PE license is required for some higher positions, it's obtained by scoring well on the two standardized assessments, and college and experience can be used to prepare for the assessments in order to gain the certification. The original sentence is a run-on and contains confusing information, so Choice *A* is incorrect. Choice *B* fixes the run-on aspect, but the sentence is indirect and awkward in construction. Choice *C* is incorrect for the same reason as Choice *B*, and it is a run on.

6. B: *Once the job begins, this line of work* is the best way to phrase this sentence. Choice *A* is incorrect because *lines* does not match up with *this*; it would instead match up with the word *these*. Choice *C* is incorrect; *these line* should say *this line*. Choice *D* is incorrect; *job begin* is faulty subject/verb agreement.

7. A: The word is spelled correctly as it is: *expertise*.

8. C: *Allows* is inappropriate because it does not stress what those in the position of aircraft engineers actually need to be able to do. *Requires* is the only alternative that fits because it actually describes necessary skills of the job.

9. D: The words *equations* and *processes* in this sentence should be plural. Choices *A, B,* and *C* have one or both words as singular, which is incorrect.

10. A: The correct response is Choice *A* because this statement's intent is to give examples as to how aircraft engineers apply mathematical equations and scientific processes towards aeronautical and aerospace issues and/or inventions. The answer is not *Therefore*, Choice *B*, or *Furthermore*, Choice *D*, because no causality is being made between ideas. Two items are neither being compared nor contrasted, so *However*, Choice *C*, is also not the correct answer.

11. A: No change is required. The comma is properly placed after the introductory phrase *In May of 2015.* Choice *B* is missing the word *In.* Choice *C* does not separate the introductory phrase from the rest of the sentence. Choice *D* places an extra comma prior to 2015.

12. A: The word *conversely* best demonstrates the opposite sentiments in this passage. Choice *B* is incorrect because it denotes agreement with the previous statement. Choice *C* is incorrect because the sentiment is not restated but opposed. Choice *D* is incorrect because the previous statement is not a cause for the sentence in question.

13. A: The correct answer is Choice *A*, *are projected*. The present tense *are* matches with the rest of the sentence. The verb *are* also matches with the plural *employment opportunities*. Choice *B* uses *will be projected*, which is incorrect because the statistic is being used as evidence, which demands a present or past tense verb. In this case it is present tense to maintain consistency. Choice *C* is incorrect because the singular verb *is* does not match with the plural subject *employment opportunities*. Choice *D* is incorrect because the past tense verb *were* does not maintain consistency with the present tense in the rest of the passage.

14. C: Choice *C* is the best answer because introductory words like "Nevertheless" are always succeeded by a comma.

15. D: The main subject and verb in this sentence are far apart from each other. The subject is *engineers* and the subject is *are projected*. Although there is a clause which interrupts the subject and the verb, they still must agree with each other.

16. D: The correct phrase should be "War on Terror." The phrase is capitalized because it was part of the campaign phrase that was launched by the U.S. government after September 11. Punctuation should always be used inside double quotes as well, making Choice *D* the best answer.

17. C: Terrorists commonly use fear and violence to achieve political goals. Choice *A* is incorrect because the subject *Terrorists* is plural while the verb *uses* is singular, so the subject and verb do not agree with each other. Choice *B* is incorrect because the word *Terrorist's* with the apostrophe *-s* shows possession, but the terrorists aren't in possession of anything in this sentence. Choice *D* is incorrect because the word *achieves* should be *achieve*.

18. A: No change is needed. Choices *B* and *C* utilize incorrect comma placements. Choice *D* utilizes an incorrect verb tense (*responding*).

19. B: The best answer Choice is *B*, *who live mainly in the South and Southwest Asia*. The directional terms *South Asia* and *Southwest Asia* are integral parts of a proper name and should therefore be capitalized.

20. D: This is the best answer choice because *America's* with the apostrophe *-s* shows possession of the word *support*, and *Israel* should be capitalized because it is a country and therefore a proper noun. Choice *A* does not show possession in the word *Americas*. Choice *B* does not capitalize the word *Israel*. Choice *C* does not show possession and does not include the necessary comma at the end of the phrase.

21. A: This sentence is correct as-is. The verb tense should be in the past—the other three answer choices either have a present or continuous verb tense, so these are incorrect.

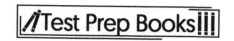

22. C: Choice *C* is the most straightforward version of this independent clause, because it follows the "subject + verb + prepositional phrase" order which usually provides the most clarity.

23. B: Although *diverging* means to separate from the main route and go in a different direction, it is used awkwardly and unconventionally in this sentence. Therefore, Choice *A* is not the answer. Choice *B* is the correct answer because it implies that the passengers distracted the terrorists, which caused a change in the plane's direction. *Converging*, Choice *C,* is incorrect because it implies that the plane met another in a central location. Although the passengers may have distracted the terrorists, they did not distract the plane. Therefore, Choice *D* is incorrect.

24. B: Choice *B* is the best answer because it is straightforward and clear. Choice *A* is incorrect because the phrase *the fact that* is redundant. Choice *C* is inverted and doesn't make much sense because the subject comes after the verb. Choice *D* is incorrect because it does not have the appropriate transition, *but*, which is intended to show a contrast to the *uncertainty* phrase that comes before it.

25. A: Choice *A* is the best choice for this sentence because it is the most straightforward and easiest to understand. Choice *B* is incorrect because it leaves out the hedging language. Choice *C* keeps the hedging language, but the sentence begins with a verb which is not the best decision for clarity. Choice *D* is incorrect because it begins with a preposition which is not the best choice for a straightforward presentation of the facts.

26. A: The word *valiantly* is spelled correctly in the original sentence.

27. D: Airspace and airports must be closed by people; they don't just close themselves, so it is proper to include an action to indicate that they were sealed off. Choice *B* is incorrect because the verb *close* is in the incorrect tense. Choice *C* is also incorrect because *airspace* does not need to become *airspaces* and the issue still remains: while there is action, it is not in the proper form to indicate human action. Choice *D* is correct because it correctly uses the helping verb *were*, which indicates human action.

28. C: This sentence contains improper verb agreement in the fragment *as cancelled flights are rescheduled. Are* is a present-tense verb while *rescheduled* is a past-tense verb. Because the attacks occurred in the past, both verbs need to be written in the past tense, as done in Choice *C*.

29. C: *Intended* means planned or meant to. *Intended* is a far better choice than *desired,* because it would communicate goals and strategy more than simply saying that Bush desired to do something. *Desired* communicates wishing or direct motive. Choices *B* and *D* have irrelevant meanings and wouldn't serve the sentence at all.

30. A: The original structure of the two sentences is correct. Choice *B* lacks the direct nature that the original sentence has. By breaking up the sentences, the connection between the Taliban's defeat and the ongoing war is separated by an unnecessary second sentence. Choice *C* corrects this problem, but the fluidity of the sentence is marred because of the awkward construction of the first sentence. Choice *D* begins well but lacks the use of *was* before *overthrown*.

31. C: Choice *C* is the best answer choice here because we have "subject + verb + prepositional phrases" which is usually the most direct sentence combination. Choice *A* is incorrect because the verbs *seeing* and *see* are repetitive. Choice *B* is incorrect because the subject and verb are separated, which is not the best way to divulge information in this particular sentence. Choice *D* is incorrect because the sentence uses a continuous verb form instead of past tense.

32. A: The comma after *result* is necessary for the sentence structure, making it an imperative component. The original sentence is correct, making Choice *A* correct. Choice *B* is incorrect because it lacks the crucial comma that introduces a new idea. Choice *C* is incorrect because a colon is unnecessary, and Choice *D* is incorrect because the addition of the word *of* is unnecessary when applied to the rest of the sentence.

33. B: Hampton was born and raised in Maywood of Chicago, Illinois in 1948. Choice *A* is incorrect because the subject and verb are at the end, and this is not the most straightforward syntax. Choice *C* is incorrect because the rest of the passage is in past tense. Choice *D* is incorrect because there should be a comma between the names of a city and a state.

34. D: Choice *D* is the best answer because it fixes the comma splice in the original sentence by creating a modifying clause at the beginning of the sentence. Choice *A* is incorrect because it contains a comma splice. Choice *B* is too wordy. Choice *C* has an unnecessary comma.

35. A: The sentence is correct as-is. Choice *B* is incorrect because we are missing the chronological signpost *after*. Choice *C* is incorrect because there is no subject represented in this sentence. Choice *D* is incorrect because it also doesn't have a subject.

36. C: Choice *C* is the best sentence because it is a compound sentence with the appropriate syntax. Choice *A* is incorrect because *resulting* is not as clear as the phrase *as a result*. Choice *B* is incorrect because the syntax is clunky—to say something happened to the NAACP rather than the NAACP making the gain is awkward. Choice *D* is incorrect because this is passive voice; active voice would be better to use here as in Choice *C*.

37. D: Choice *D* is the best answer. Choice *A* is incorrect; if the second clause was independent we would need a comma in the middle, but the two clauses have the same single subject, so a comma is unnecessary. Choice *B* is incorrect; semicolons go in between two independent clauses without conjunctions. Choice *C* is incorrect; there needs to be a conjunction before the word *relocated*.

38. B: *enabled him to quickly rise* is the best answer. Although the passage is in past tense, the verb *enabled* is followed by the "to + infinitive." The infinitive is the base form of the verb; that's why *rise* is the best answer choice, because the verb follows *to* and represents the infinitive after the verb *enabled*.

39. A: No change is needed. The list of events should be separated by a comma. Choice *B* is incorrect. Although a colon can be used to introduce a list of items, it is not a conventional choice for separating items within a series. Semicolons are used to separate at least three items in a series that have an internal comma. Therefore, Choice *C* is incorrect. Choice *D* is incorrect because a dash is not a conventional choice for punctuating items in a series.

40. C: *Hampton's greatest achievement as the leader of the BPP* is the best way to phrase this sentence because it is the most straightforward and clear. Choice *A* is incorrect because the sentence starts out with a gerund when it could start out with Hampton, reducing confusion as to who is leading the BPP. Choice *B* is incorrect for the same reason; the sentence starts off with the subject inside the preposition, which is not the best syntax. Choice *D* is incorrect because it doesn't signify that the achievement was Hampton's.

41. D: Choice *D* is the most straightforward answer, "Hampton held a press conference." Choice *A* is incorrect because usually one isn't held "by" a press conference. Choice *B* is incorrect because likewise, a press conference doesn't "hold" someone, but someone would "hold" a press conference. Choice *C* is incorrect because again, someone isn't held "by" a press conference, and the syntax is inverted here.

42. A: The term *neutralize* means to counteract, or render ineffective, which is exactly what the FBI is wanting to do. Accommodate means to be helpful or lend aid, which is the opposite of *neutralize*. Therefore, Choice *B* is incorrect. *Assuage* means to ease, while *praise* means to express warm feeling, so they are in no way close to the needed context. Therefore, *neutralize* is the best option, making Choice *A* the correct answer.

43. C: Choice *C* is the most straightforward answer choice. Choice *A* is incorrect because the sentence is inverted and not very clear. Choice *B* is incorrect because the change in syntax changes the meaning in the sentence. Choice *D* isn't the best way to organize the sentence because it creates wordiness.

44. D: Choice *D* is the best answer choice because it matches with the past tense of the passage. The other answer choices are either in present tense or a present continuous tense.

45. B: Choice *B* is the most straightforward sentence, beginning with the subject and verb, then leading off with two prepositions. Choice *A* is incorrect because the sentence is passive and thus more wordy than the active voice in Choice *B*. Choice *C* is incorrect because the sentence starts out with two prepositions which isn't the best syntax for clarity here. Choice *D* is incorrect because the verb is in present tense instead of past tense.

46. A: Choice *A* is the best answer because *commemorates* means to show respect for something through recognition. Choice *B*, *disparages,* means to discourage, so this is incorrect. Choice *C*, *exculpates*, means to show that someone is not guilty of wrongdoing, and this does not make sense within the context of the sentence. Choice *D*, *brandish*, means to swing or flaunt, and is not the best choice here.

47. D: Choice *D* is the best answer. Choice *A* is incorrect; the comma is not necessary here because both clauses use the same subject, *Floods*. Choice *B* is incorrect because *causing by* is not an appropriate verb phrase. Choice *C* is incorrect because *so to cause by* is also not an appropriate verb phrase.

48. C: The sentence is best written with a comma before the clause *as are low-lying regions*. Choice *A* is incorrect because the comma before the clause is missing. Choice *B* is incorrect; there is a comma splice and this sentence is repetitive. Choice *D* is incorrect because the syntax is not straightforward, and it also is repetitive.

49. D: Choice *D* is the best arrangement of the sentence. Choice *A* is incorrect; *sun* should be possessive of *radiation* and thus have an apostrophe -*s* at the end of the word. Choice *B* is incorrect; the addition of the word *which* adds unnecessary wordiness to the sentence. Choices *C* is incorrect; the word *being* is used twice which creates repetition.

50. B: In this sentence, the word *ocean* does not require an *s* after it to make it plural because *ocean levels* is plural. Therefore, Choices *A* and *C* are incorrect. Because the sentence is referring to multiple— if not all ocean levels—*ocean* does not require an apostrophe ('*s*) because that would indicate that only one ocean is the focus, which is not the case. Choice *D* does not fit well into the sentence and, once again, we see that *ocean* has an *s* after it. This leaves Choice *B,* which correctly completes the sentence and maintains the intended meaning.

51. C: Choice *C* is the best answer. Choice *A* is incorrect because *the water* is doing the overflowing, so it should come before the phrase *to overflow*. Choice *B* is incorrect because *may been causing* isn't an appropriate verb phrase. Choice *D* is incorrect because the word *water* is used repetitively.

52. D: Choice *A* is incorrect; semicolons separate items in a list only when an item in the list contains a comma itself. Choice *B* is incorrect; items in a list should always be separated by commas. Choice *C* is incorrect because the word *beavers* is missing a possessive -'s at the end.

53. B: The best conjunction to use in this sentence is the word *or*. The other conjunctions used, *then, so,* and *but*, do not fit in the context of the sentence here.

54. A: Choice *A* is the best answer. Choice *B* is incorrect because it has incorrect subject/verb agreement, *waters begins*. Choice *C* is incorrect because it is missing the word *to* in front of the word *grow*. Choice *D* is incorrect; technically this sentence is grammatically correct, but by using the conjunction *and,* it implies that water begins to drain and mold grows at the same time, which is not what the passage intended.

55. A: Choice *C* can be eliminated because creating a new sentence with *not* is grammatically incorrect, and it throws off the rest of the sentence. Choice *B* is incorrect because a comma is needed after *devastation* in the sentence. Choice *D* is also incorrect because *while* is a poor substitute for *although*. *Although* in this context is meant to show contradiction with the idea that floods are associated with devastation. Therefore, none of these choices would be suitable revisions because the original was correct.

56. B: Choice *B* is the best answer because it is the most straightforward syntax. Choice *A* is incorrect because *floodplains of rivers* is unnecessarily long. Choice *C* is incorrect because a comma is missing after the prepositional phrase. Choice *D* is incorrect because the word *thousands* is plural and therefore should not have an apostrophe.

57. B: Choice *B* is the best answer here. Choice *A* is incorrect because the text should specify what type of examples are about to be named. Choice *C* is incorrect because items in a list should be separated by commas, not by ellipses. Choice *D* is incorrect because, again, items in a list should be separated by commas.

58. D: Choice *D* is correct because it is in the indefinite present tense and is consistent with the passage. Choice *A* is incorrect because *caused* is inconsistent with the rest of the verb tense in the passage. Choice *B* is incorrect because this does not agree with the subject *the flooding*. *Causing,* Choice *C,* is in the present continuous tense which does not fit with the subject, *Flooding . . . causing.*

59. A: To *project* means to anticipate or forecast. This goes very well with the sentence because it describes how new technology is trying to estimate flood activity in order to prevent damage and save lives. *Project* in this case needs to be assisted by *to* in order to function in the sentence. Therefore, Choice *A* is correct. Choices *B* and *D* are the incorrect tenses. Choice *C* is incorrect because it lacks *to*.

60. B: The correct conjunction is *then*. The word *then* implies *after that* or *afterward*.

61. C: Choice *C* is the correct answer. Shorter poems should be in quotation marks, but tiles of long, epic poems like *The Odyssey* should be written with italics.

62. A: Choice *A* is the best answer. *However* is an appropriate word to begin this sentence since it illustrates the idea of contrast: some of the females Odysseus encountered were helpful, *however* some were clearly not. Therefore, we can eliminate Choice *D*. Choice *B* is also incorrect because it eliminates the word *several*, which is also useful in distinguishing that, while some female characters were benevolent, several were deceptive and even harmful. Choice *C* is incorrect because it lacks the comma after *However*.

63. D: Choice *D* is the best answer because it uses a comma between an independent clause and a dependent clause and gets rid of the word *himself*, eliminating wordiness. Choice *A* is incorrect because the comma between two independent clauses causes a comma splice. Additionally, the word *himself* is repetitive. Choice *B* is incorrect because the absence of a semicolon or period after *harder* creates a run-on sentence.

64. C: We would say *By the time Odysseus reaches* home to denote that one event occurs before the other. The other answer choices don't make this distinction.

65. A: Choice *A* is correct because it is in active voice rather than passive voice. Choice *B* is incorrect because *he had to be concealed himself* is passive voice and also creates wordiness with *himself*. Choice *C* makes the sentence grammatically incorrect. Choice *D* is incorrect because *concealing* does not go with the helping verb *had*. *Had concealing* is not a proper verb phrase.

66. B: Choice *B* is the best answer. Choice *A* is incorrect because it is a run-on sentence. Choice *C* is incorrect because the comma between two independent clauses creates a comma splice. Choice *D* is incorrect; semicolons should have independent clauses on either side of them, and the first phrase is not a complete sentence.

67. B: Choice *B*, *conceals,* is the best option because Odysseus does in fact hide (or conceal) his true identity behind a false name. Choice *A* is incorrect; *relinquishes* means to give up or to voluntarily cease control of something. This is not something Odysseus does because he does not surrender his true name—he just hides it. He does not *withhold* (Choice *C*) his true identity but offers an alternative. Choice *D*, *surrenders,* is synonymous with *relinquishes*.

68. A: Reading through the sentence, one can see that it flows and uses proper punctuation and grammar, which means that no change is necessary; Choice *A* is the correct answer. Choice *B* lacks the necessary comma after *Clearly*. Choice *C* utilizes an unnecessary semicolon. Choice *D* makes the sentence awkward by placing *clearly* in the middle of the sentence.

69. C: Choice *C* is the best answer choice. Choice *A* is incorrect because the author of the passage uses present tense when speaking about the text in question, so the verb should be *describes*. Choice *B* is incorrect because *passages* should be *passage*, singular. Choice *D* is incorrect because the word *These* should be *This*, describing a singular passage.

70. B: Choice *B* is the best answer because it avoids the wordiness of the original phrase *also like*, which makes Choice *A* incorrect. Choice *C* is compelling, but it lacks the necessary information that the original sentence has: we need to have Polyphemus still in the sentence. Choice *D* is a decent option, but much of it is unnecessary and already addressed in the remainder of the sentence that isn't underlined.

71. B: The issue with this word is that it lacks proper punctuation. *One* is referring to an individual, and it needs to show possession of *identity*. Therefore, *ones* must have an apostrophe to show ownership and to be correct. Thus, Choice *B*, *One's,* is the correct answer.

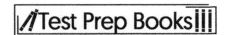

72. C: The current phrase repeats itself, so we can eliminate Choice *A*. Choices *B* and *D* are compelling but are somewhat specific. What must be communicated is that all men who reach the island, in whatever fashion, risk being seduced by Circe. Choice *C* is the most practical revision and is more direct, while allowing for other circumstances in which men come to Circe's island.

73. B: The best answer is Choice *B*, *ensure*, which means to make certain. Odysseus is trying to make certain that he and the crew will be safe by beating Circe at her own game. Choice *A*, *assure*, means to speak to someone in a way that eliminates doubt. This indirectly relates to what Odysseus wants to do, but to *assure* specifically refers to speaking to someone. Choices *C* and *D* are totally irrelevant. *Vindicate* means to free from blame, while *prevent* means to stop from happening.

74. B: The best verb to use here is *returns*. When talking about text we use present tense verbs.

75. A: Choice *A* is the best answer. Choice *B* is incorrect because it is missing a comma after the introductory clause, *As the text progresses*. Choice *C* is incorrect; Odysseus is the subject of the sentence and must be outside of any phrases—Choice *C* has the subject *Odysseus* inside a prepositional phrase starting with the word *with*, so this is incorrect. Choice *D* is incorrect because again, our subject is inside a clause and therefore does not pair with the verb *encounters*.

Math

1. D: This problem can be solved using basic arithmetic. Xavier starts with 20 apples, then gives his sister half, so 20 divided by 2.

$$\frac{20}{2} = 10$$

He then gives his neighbor 6, so 6 is subtracted from 10.

$$10 - 6 = 4$$

Lastly, he uses ¾ of his apples to make an apple pie, so to find remaining apples, the first step is to subtract ¾ from one and then multiply the difference by 4.

$$\left(1 - \frac{3}{4}\right) \times 4 = ?$$

$$\left(\frac{4}{4} - \frac{3}{4}\right) \times 4 = ?$$

$$\left(\frac{1}{4}\right) \times 4 = 1$$

2. C: The product of two irrational numbers can be rational or irrational. Sometimes, the irrational parts of the two numbers cancel each other out, leaving a rational number. For example, $\sqrt{2} \times \sqrt{2} = 2$ because the roots cancel each other out. Technically, the product of two irrational numbers can be complex because complex numbers can have either the real or imaginary part (in this case, the imaginary part) equal zero and still be considered a complex number. However, Choice *D* is incorrect because the product of two irrational numbers is not an imaginary number so saying the product is complex *and* imaginary is incorrect.

3. B: The car is traveling at a speed of five meters per second. On the interval from one to three seconds, the position changes by ten meters. By making this change in position over time into a rate, the speed becomes ten meters in two seconds or five meters in one second.

4. C: The number negative four is classified as a real number because it exists and is not imaginary. It is rational because it does not have a decimal that never ends. It is an integer because it does not have a fractional component. The next classification would be whole numbers, for which negative four does not qualify because it is negative. Choices *D* and *E* are wrong because -4 is not considered an irrational number because it does not have a never-ending decimal component.

5. B: Since $850 is the price *after* a 20% discount, $850 represents 80% of the original price. To determine the original price, set up a proportion with the ratio of the sale price (850) to original price (unknown) equal to the ratio of sale percentage (where x represents the unknown original price):

$$\frac{850}{x} = \frac{80}{100}$$

To solve a proportion, cross multiply the numerators and denominators and set the products equal to each other: $(850)(100) = (80)(x)$. Multiplying each side results in the equation $85,000 = 80x$.

To solve for x, divide both sides by 80: $\frac{85,000}{80} = \frac{80x}{80}$, resulting in $x = 1062.5$. Remember that x represents the original price. Subtracting the sale price from the original price ($\$1062.50 - \$850$) indicates that Frank saved $212.50.

6. D: $0.45

List the givens.

$$Store\ coffee = \$1.23/lbs$$

$$Local\ roaster\ coffee = \$1.98/1.5\ lbs$$

Calculate the cost for 5 lbs of store brand.

$$\frac{\$1.23}{1\ lbs} \times 5\ lbs = \$6.15$$

Calculate the cost for 5 lbs of the local roaster.

$$\frac{\$1.98}{1.5\ lbs} \times 5\ lbs = \$6.60$$

Subtract to find the difference in price for 5 lbs.

$$\begin{array}{r} \$6.60 \\ -\$6.15 \\ \hline \$0.45 \end{array}$$

7. C: The first step in solving this problem is expressing the result in fraction form. Separate this problem first by solving the division operation of the last two fractions. When dividing one fraction by another, invert or flip the second fraction and then multiply the numerator and denominator.

$$\frac{7}{10} \times \frac{2}{1} = \frac{14}{10}$$

Next, multiply the first fraction with this value:

$$\frac{3}{5} \times \frac{14}{10} = \frac{42}{50}$$

In this instance, you can find the decimal form by converting the fraction into $\frac{x}{100}$, where x is the number from which the final decimal is found. Multiply both the numerator and denominator by 2 to get the fraction as an expression of $\frac{x}{100}$.

$$\frac{42}{50} \times \frac{2}{2} = \frac{84}{100}$$

In decimal form, this would be expressed as 0.84.

8. E: To find the average of a set of values, add the values together and then divide by the total number of values. In this case, include the unknown value of what Dwayne needs to score on his next test, in order to solve it.

$$\frac{78 + 92 + 83 + 97 + x}{5} = 90$$

Add the unknown value to the new average total, which is 5. Then multiply each side by 5 to simplify the equation, resulting in:

$$78 + 92 + 83 + 97 + x = 450$$

$$350 + x = 450$$

$$x = 100$$

Dwayne would need to get a perfect score of 100 in order to get an average of at least 90.

Test this answer by substituting back into the original formula:

$$\frac{78 + 92 + 83 + 97 + 100}{5} = 90$$

9. D: 1, 2, 3, 4, 6, 12. A given number divides evenly by each of its factors to produce an integer (no decimals). The number 5, 7, 8, 9, 10, 11 (and their opposites) do not divide evenly into 12. Therefore, these numbers are not factors.

10. E: The augmented matrix that represents the system of equations has dimensions 4 × 3 because there are three equations with three unknowns. The coefficients of the variables make up the first three columns, and the last column is made up of the numbers to the right of the equals sign. This system can be solved by reducing the matrix to row-echelon form, where the last column gives the solution for the unknown variables.

11. B: There are two zeros for the function: $x = 0, -2$.

The zeros can be found several ways, but this particular equation can be factored into:

$$f(x) = x(x^2 + 4x + 4) = x(x + 2)(x + 2)$$

By setting each factor equal to zero and solving for x, there are two solutions. On a graph these zeros can be seen where the line crosses the x-axis.

12. D: This problem involves a composition function, where one function is plugged into the other function. In this case, the $f(x)$ function is plugged into the $g(x)$ function for each x-value. The composition equation becomes:

$$g\big(f(x)\big) = 2^3 - 3(2^2) - 2(2) + 6$$

Simplifying the equation gives the answer:

$$g\big(f(x)\big) = 8 - 3(4) - 2(2) + 6 = 8 - 12 - 4 + 6 = -2$$

13. D: This system of equations involves one quadratic function and one linear function, as seen from the degree of each equation. One way to solve this is through substitution. Solving for y in the second equation yields $y = x + 2$. Plugging this equation in for the y of the quadratic equation yields:

$$x^2 - 2x + x + 2 = 8$$

Simplifying the equation, it becomes:

$$x^2 - x + 2 = 8$$

Setting this equal to zero and factoring, it becomes:

$$x^2 - x - 6 = 0 = (x - 3)(x + 2)$$

Solving these two factors for x gives the zeros $x = 3, -2$. To find the y-value for the point, each number can be plugged in to either original equation. Solving each one for y yields the points $(3, 5)$ and $(-2, 0)$.

14. D: The expression is simplified by collecting like terms. Terms with the same variable and exponent are like terms, and their coefficients can be added.

15. A: Finding the product means distributing one polynomial onto the other. Each term in the first must be multiplied by each term in the second. Then, like terms can be collected. Multiplying the factors yields the expression:

$$20x^3 + 4x^2 + 24x - 40x^2 - 8x - 48$$

Collecting like terms means adding the x^2 terms and adding the x terms. The final answer after simplifying the expression is:

$$20x^3 - 36x^2 + 16x - 48$$

16. E: Finding the zeros for a function by factoring is done by setting the equation equal to zero, then completely factoring. Since there was a common x for each term in the provided equation, that would be factored out first. Then the quadratic that was left could be factored into two binomials, which are $(x + 1)(x - 4)$. Setting each factor equal to zero and solving for x yields three zeros.

17. A: 13 nurses

Using the given information of 1 nurse to 25 patients and 325 patients, set up an equation to solve for number of nurses (N):

$$\frac{N}{325} = \frac{1}{25}$$

Multiply both sides by 325 to get N by itself on one side.

$$\frac{N}{1} = \frac{325}{25} = 13 \; nurses$$

18. E: The equation can be solved by factoring the numerator into $(x + 6)(x - 5)$.

Since that same factor exists on top and bottom, that factor $(x - 5)$ cancels.

This leaves the equation $x + 6 = 11$.

Solving the equation gives the answer $x = 5$. When this value is plugged into the equation, it yields a zero in the denominator of the fraction. Since this is undefined, there is no solution.

19. D: This problem can be solved by using unit conversion. The initial units are miles per minute. The final units need to be feet per second. Converting miles to feet uses the equivalence statement 1 mile equals 5,280 feet. Converting minutes to seconds uses the equivalence statement 1 minute equals 60 seconds. Setting up the ratios to convert the units is shown in the following equation:

$$\frac{72 \; mi}{90 \; min} \times \frac{1 \; min}{60 \; s} \times \frac{5280 \; ft}{1 \; mi} = 70.4 \frac{ft}{s}$$

The initial units cancel out, and the new units are left.

20. B: The formula can be manipulated by dividing both the length, *l*, and the width, *w*, on both sides. The length and width will cancel on the right, leaving height by itself.

21. B: The domain is all possible input values, or *x*-values. For this equation, the domain is every number greater than or equal to zero. There are no negative numbers in the domain because taking the square root of a negative number results in an imaginary number.

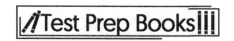
22. D: This problem can be solved by setting up a proportion involving the given information and the unknown value. The proportion is:

$$\frac{21\ pages}{4\ nights} = \frac{140\ pages}{x\ nights}$$

Solving the proportion by cross-multiplying, the equation becomes $21x = 4 \times 140$, where $x = 26.67$. Since it is not an exact number of nights, the answer is rounded up to 27 nights. Twenty-six nights would not give Sarah enough time.

23. D: The slope from this equation is 50, and it is interpreted as the cost per gigabyte used. Since the g-value represents number of gigabytes and the equation is set equal to the cost in dollars, the slope relates these two values. For every gigabyte used on the phone, the bill goes up 50 dollars.

24. C: Graphing the function $y = \cos(x)$ shows that the curve starts at $(0, 1)$, has an amplitude of 2, and a period of 2π. This same curve can be constructed using the sine graph, by shifting the graph to the left $\frac{\pi}{2}$ units. This equation is in the form:

$$y = \sin\left(x + \frac{\pi}{2}\right)$$

25. B: Start by squaring both sides to get $1 + x = 16$. Then subtract 1 from both sides to get $x = 15$.

26. A: This inverse of a function is found by switching the x and y in the equation and solving for y. In the given equation, solving for y is done by adding 5 to both sides, then dividing both sides by 3. This answer can be checked on the graph by verifying the lines are reflected over $y = x$.

27. B: The zeros of this function can be found by using the quadratic formula:

$$x = \frac{-b \pm \sqrt{b^2 - 4ac}}{2a}$$

Identifying a, b, and c can also be done from the equation because it is in standard form. The formula becomes:

$$x = \frac{0 \pm \sqrt{0^2 - 4(1)(4)}}{2(1)} = \frac{\sqrt{-16}}{2}$$

Since there is a negative underneath the radical, the answer is a complex number:

$$x = \pm 2i$$

28. E: Setting up a proportion is the easiest way to represent this situation. The proportion becomes $\frac{20}{x} = \frac{40}{100}$, where cross-multiplication can be used to solve for x. The answer can also be found by observing the two fractions as equivalent, knowing that twenty is half of forty, and fifty is half of one-hundred.

29. C: Division with scientific notation can be solved by grouping the first terms together and grouping the tens together. The first terms can be divided, and the tens terms can be simplified using the rules for exponents. The initial expression becomes $0.4 * 10^4$. This is not in scientific notation because the first

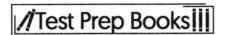

number is not between 1 and 10. Shifting the decimal and subtracting one from the exponent yields $4.0 * 10^3$.

30. A: The relative error can be found by finding the absolute error and making it a percent of the true value. The absolute error is $36 - 35.75 = 0.25$. This error is then divided by 36—the true value—to find 0.7%.

31. B: The y-intercept of an equation is found where the x-value is zero. Plugging zero into the equation for x allows the first two terms to cancel out, leaving -4.

32. A: The equation is *even* because:

$$f(-x) = f(x)$$

Plugging in a negative value will result in the same answer as when plugging in the positive of that same value. The function:

$$f(-2) = \frac{1}{2}(-2)^4 + 2(-2)^2 - 6 = 8 + 8 - 6 = 10$$

yields the same value as:

$$f(2) = \frac{1}{2}(2)^4 + 2(2)^2 - 6 = 8 + 8 - 6 = 10$$

33. C: The equation $x = 3$ is not a function because it does not pass the vertical line test. This test is made from the definition of a function, where each x-value must be mapped to one, and only one, y-value. This equation is a vertical line, so the x-value of 3 is mapped with an infinite number of y-values.

34. A: The function can be factored to identify the zeros. First, the term $3x$ is factored out to the front because each term contains $3x$. Then, the quadratic is factored into $(x + 3)(x - 2)$.

35. C: A die has an equal chance for each outcome. Since it has six sides, each outcome has a probability of $\frac{1}{6}$. The chance of a 1 or a 2 is therefore $\frac{1}{6} + \frac{1}{6} = \frac{1}{3}$.

36. A: The slope is given by:

$$m = \frac{y_2 - y_1}{x_2 - x_1} = \frac{0 - 4}{0 - (-3)} = -\frac{4}{3}$$

37. B: The table shows values that are increasing exponentially. The differences between the inputs are the same, while the differences in the outputs are changing by a factor of 2. The values in the table can be modeled by the equation $f(x) = 2^x$.

38. C: The formula for continually compounded interest is $A = Pe^{rt}$. Plugging in the given values to find the total amount in the account yields the equation:

$$A = 2000e^{0.05*8} = 2983.65$$

39. E: The sample space is made up of $8 + 7 + 6 + 5 = 26$ balls. The probability of pulling each individual ball is $^1/_{26}$. Since there are 7 yellow balls, the probability of pulling a yellow ball is $^7/_{26}$.

40. B: For the first card drawn, the probability of a King being pulled is $\frac{4}{52}$. Since this card isn't replaced, if a King is drawn first the probability of a King being drawn second is $\frac{3}{51}$. The probability of a King being drawn in both the first and second draw is the product of the two probabilities:

$$\frac{4}{52} \times \frac{3}{51} = \frac{12}{2652}$$

This fraction, when divided by 12, equals $\frac{1}{221}$.

41. E: The addition rule is necessary to determine the probability because a 6 can be rolled on either roll of the die. The rule used is:

$$P(A \text{ or } B) = P(A) + P(B) - P(A \text{ and } B)$$

The probability of a 6 being individually rolled is $\frac{1}{6}$ and the probability of a 6 being rolled twice is:

$$\frac{1}{6} \times \frac{1}{6} = \frac{1}{36}$$

Therefore, the probability that a 6 is rolled at least once is:

$$\frac{1}{6} + \frac{1}{6} - \frac{1}{36} = \frac{11}{36}$$

42. A: $(A \cap B)$ is equal to the intersection of the two sets A and B, which is $\{1, 2, 3, 4, 5\}$. $A - (A \cap B)$ is equal to the elements of A that are *not* included in the set $(A \cap B)$. Therefore:

$$A - (A \cap B) = \{6, 7, 8, 9, 10\}$$

43. B: An equilateral triangle has three sides of equal length, so if the total perimeter is 18 feet, each side must be 6 feet long. A square with sides of 6 feet will have an area of $6^2 = 36$ square feet.

44. A: If each man gains 10 pounds, every original data point will increase by 10 pounds. Therefore, the man with the original median will still have the median value, but that value will increase by 10.

The smallest value and largest value will also increase by 10 and, therefore, the difference between the two won't change. The range does not change in value and, thus, remains the same.

45. C: Because the triangles are similar, the lengths of the corresponding sides are proportional. Therefore:

$$\frac{30 + x}{30} = \frac{22}{14} = \frac{y + 15}{y}$$

This results in the equation $14(30 + x) = 22 \times 30$ which, when solved, gives $x = 17.1$.

The proportion also results in the equation $14(y + 15) = 22y$ which, when solved, gives $y = 26.3$.

46. A: Robert accomplished his task on Tuesday in ¾ the time compared to Monday. He must have worked 4/3 as fast.

47. B: To simplify this inequality, subtract 3 from both sides to get $-\frac{1}{2}x \geq -1$. Then, multiply both sides by -2 (remembering this flips the direction of the inequality) to get $x \leq 2$.

48. E: There are two ways to approach this problem. Each value can be substituted into each equation. Choice *A* can be eliminated, since:

$$4^2 + 16 = 32$$

Choice *B* can be eliminated, since:

$$4^2 + 4 \cdot 4 - 4 = 28$$

Choice *C* can be eliminated, since:

$$4^2 - 2 \cdot 4 - 2 = 6$$

But, plugging in either value into $x^2 - 16$ gives:

$$(\pm 4)^2 - 16 = 16 - 16 = 0$$

48. E: There are two ways to approach this problem. Each value can be substituted into each equation. Choice *A* can be eliminated, since:

$$4^2 + 16 = 32$$

Choice *B* can be eliminated, since:

$$4^2 + 4 \cdot 4 - 4 = 28$$

Choice *C* can be eliminated, since:

$$4^2 - 2 \cdot 4 - 2 = 6$$

But, plugging in either value into $x^2 - 16$ gives:

$$(\pm 4)^2 - 16 = 16 - 16 = 0$$

49. B: The technique of completing the square must be used to change:

$$4x^2 + 4y^2 - 16x - 24y + 51 = 0$$

into the standard equation of a circle. First, the constant must be moved to the right-hand side of the equals sign, and each term must be divided by the coefficient of the x^2 term (which is 4). The x and y terms must be grouped together to obtain:

$$x^2 - 4x + y^2 - 6y = -\frac{51}{4}$$

Next, the process of completing the square must be completed for each variable. This gives:

$$(x^2 - 4x + 4) + (y^2 - 6y + 9) = -\frac{51}{4} + 4 + 9$$

The equation can be written as:

$$(x - 2)^2 + (y - 3)^2 = \frac{1}{4}$$

Therefore, the center of the circle is (2, 3) and the radius is:

$$\sqrt{1/4} = 1/2$$

50. D: When an ordered pair is reflected over an axis, the sign of one of the coordinates must change. When it's reflected over the x-axis, the sign of the y-coordinate must change. The x-value remains the same. Therefore, the new ordered pair is $(-3, 4)$.

51. A: Because the volume of the given sphere is 288π cubic meters, this means $\frac{4}{3}\pi r^3 = 288\pi$. This equation is solved for r to obtain a radius of 6 meters. The formula for the surface area of a sphere is $4\pi r^2$, so if $r = 6$ in this formula, the surface area is 144π square meters.

52. E: SOHCAHTOA is used to find the missing side length. Because the angle and adjacent side are known, $\tan 60 = \frac{x}{13}$. Making sure to evaluate tangent with an argument in degrees, this equation gives:

$$x = 13 \tan 60 = 13 \times \sqrt{3} = 22.52$$

53. A: Let the unknown score be x. The average will be:

$$\frac{5 \times 50 + 4 \times 70 + x}{10} = \frac{530 + x}{10} = 55$$

Multiply both sides by 10 to get $530 + x = 550$, or $x = 20$.

54. A: First, the sample mean must be calculated:

$$\bar{x} = \frac{1}{4}(1 + 3 + 5 + 7) = 4$$

The standard deviation of the data set is $s = \sqrt{\frac{\Sigma(x-\bar{x})^2}{n-1}}$, and $n = 4$ represents the number of data points. Therefore:

$$s = \sqrt{\frac{1}{3}[(1 - 4)^2 + (3 - 4)^2 + (5 - 4)^2 + (7 - 4)^2]} = \sqrt{\frac{1}{3}(9 + 1 + 1 + 9)} = 2.58$$

55. D: For manufacturing costs, there is a linear relationship between the cost to the company and the number produced, with a y-intercept given by the base cost of acquiring the means of production, and a slope given by the cost to produce one unit. In this case, that base cost is $50,000, while the cost per unit is $40. So, $y = 40x + 50,000$.

56. C: The expected value is equal to the total sum of each product of individual score and probability. There are 36 possible rolls.

The probability of rolling a 2 is $\frac{1}{36}$.

The probability of rolling a 3 $\frac{2}{36}$.

The probability of rolling a 4 is $\frac{3}{36}$.

The probability of rolling a 5 is $\frac{4}{36}$.

The probability of rolling a 6 is $\frac{5}{36}$.

The probability of rolling a 7 is $\frac{6}{36}$.

The probability of rolling an 8 is $\frac{5}{36}$.

The probability of rolling a 9 is $\frac{4}{36}$.

The probability of rolling a 10 is $\frac{3}{36}$.

The probability of rolling an 11 is $\frac{2}{36}$.

Finally, the probability of rolling a 12 is $\frac{1}{36}$.

Each possible outcome is multiplied by the probability of it occurring. Like this:

$$2 \times \frac{1}{36} = a$$

$$3 \times \frac{2}{36} = b$$

$$4 \times \frac{3}{36} = c$$

And so forth.

Then all of those results are added together:

$$a + b + c \ldots = expected\ value$$

In this case, it equals 7.

57. B: From the slope-intercept form, $y = mx + b$, it is known that b is the y-intercept, which is 1. Compute the slope as $\frac{2-1}{1-0} = 1$, so the equation should be $y = x + 1$.

58. A: This has the form $t^2 - y^2$, with $t = x^2$ and $y = 4$.

It's also known that:

$$t^2 - y^2 = (t + y)(t - y)$$

and substituting the values for *t* and *y* into the right-hand side gives $(x^2 - 4)(x^2 + 4)$.

59. A: Simplify this to:

$$(4x^2y^4)^{\frac{3}{2}} = 4^{\frac{3}{2}}(x^2)^{\frac{3}{2}}(y^4)^{\frac{3}{2}}$$

Now:

$$4^{\frac{3}{2}} = (\sqrt{4})^3 = 2^3 = 8$$

For the other, recall that the exponents must be multiplied, so this yields:

$$8x^{2 \cdot \frac{3}{2}}y^{4 \cdot \frac{3}{2}} = 8x^3y^6$$

60. B: The independent variable's coordinate at the vertex of a parabola (which is the highest point, when the coefficient of the squared independent variable is negative) is given by $x = -\frac{b}{2a}$. Substitute and solve for *x* to get:

$$x = -\frac{4}{2(-16)} = \frac{1}{8}$$

Using this value of *x*, the maximum height of the ball (*y*), can be calculated.

Substituting *x* into the equation yields:

$$h(t) = -16\frac{1^2}{8} + 4\frac{1}{8} + 6 = 6.25$$

Reading

Fiction

1. B: Strong dislike. This vocabulary question can be answered using context clues. Based on the rest of the conversation, the reader can gather that Albert isn't looking forward to his marriage. As the Count notes that "you don't appear to me to be very enthusiastic on the subject of this marriage," and also remarks on Albert's "objection to a young lady who is both rich and beautiful," readers can guess Albert's feelings. The answer choice that most closely matches "objection" and "not . . . very enthusiastic" is B, *strong dislike*.

2. C: Their name is more respected than the Danglars'. This inference question can be answered by eliminating incorrect answers. Choice *A* is tempting, considering that Albert mentions money as a concern in his marriage. However, although he may not be as rich as his fiancée, his father still has a stable income of 50,000 francs a year. Choice *B* isn't mentioned at all in the passage, so it's impossible to make an inference. Finally, Choice *D* is clearly false because Albert's father arranged his marriage, but his mother doesn't approve of it. Evidence for Choice *C* can be found in the Count's comparison of

Albert and Eugénie: "she will enrich you, and you will ennoble her." In other words, the Danglars are wealthier but the Morcerf family has a more noble background.

3. D: Apprehensive. There are many clues in the passage that indicate Albert's attitude towards his marriage—far from enthusiastic, he has many reservations. This question requires test takers to understand the vocabulary in the answer choices. *Pragmatic* is closest in meaning to *realistic*, and *indifferent* means *uninterested*. The only word related to feeling worried, uncertain, or unfavorable about the future is *apprehensive*.

4. B: He is like a wise uncle, giving practical advice to Albert. Choice *A* is incorrect because the Count's tone is friendly and conversational. Choice *C* is also incorrect because the Count questions why Albert doesn't want to marry a young, beautiful, and rich girl. While the Count asks many questions, he isn't particularly *probing* or *suspicious*—instead, he's asking to find out more about Albert's situation and then give him advice about marriage.

5. A: She belongs to a noble family. Though Albert's mother doesn't appear in the scene, there's more than enough information to answer this question. More than once is his family's noble background mentioned (not to mention that Albert's mother is the Comtesse de Morcerf, a noble title). The other answer choices can be eliminated—she is deeply concerned about her son's future; money isn't her highest priority because otherwise she would favor a marriage with the wealthy Danglars; and Albert describes her "clear and penetrating judgment," meaning she makes good decisions.

6. C: The richest people in society were also the most respected. The Danglars family is wealthier but the Morcerf family has a more aristocratic name, which gives them a higher social standing. Evidence for the other answer choices can be found throughout the passage: Albert mentioned receiving money from his father's fortune after his marriage; Albert's father has arranged this marriage for him; and the Count speculates that Albert's mother disapproves of this marriage because Eugénie isn't from a noble background like the Morcerf family, implying that she would prefer a match with a girl from aristocratic society.

7. A: He seems reluctant to marry Eugénie, despite her wealth and beauty. This is a reading comprehension question, and the answer can be found in the following lines: "'I confess,' observed Monte Cristo, 'that I have some difficulty in comprehending your objection to a young lady who is both rich and beautiful.'" Choice *B* is the opposite (Albert's father is the one who insists on the marriage), Choice *C* incorrectly represents Albert's eagerness to marry, and Choice *D* describes a more positive attitude than Albert actually feels (*repugnance*).

8. B: The passage is mostly made up of dialogue. We can see this in the way the two characters communicate with each other, in this case through the use of quotations marks, or dialogue. Narration is when the narrator (not the characters) is explaining things that are happening in the story. Description is when the narrator describes a specific setting and its images. Explanation is when the author is analyzing or defining something for the reader's benefit.

9. D: The meaning of the word *ennoble* in the middle of the paragraph means to give someone a noble rank or title. In the passage, we can infer that Albert is noble but not rich, and Mademoiselle Eugénie is rich but not noble.

10. C: Because he finally understands Albert's point of view but still doesn't agree with it. The other choices aren't mentioned anywhere in the passage. Remember, although this passage is part of a larger

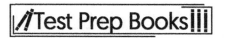

text, the test taker should only pay attention to what's in the passage itself in order to find the correct answer.

Social Science

11. C: In lines 6 and 7, it is stated that avarice can prevent a man from being necessitously poor, but too timorous, or fearful, to achieve real wealth. According to the passage, avarice does not tend to make a person very wealthy. The passage states that oppression, not avarice, is the consequence of wealth. The passage does not state that avarice drives a person's desire to be wealthy.

12. D: Paine believes that the distinction that is beyond a natural or religious reason is between king and subjects. He states that the distinction between good and bad is made in heaven. The distinction between male and female is natural. He does not mention anything about the distinction between humans and animals.

13. A: The passage states that the Heathens were the first to introduce government by kings into the world. The quiet lives of patriarchs came before the Heathens introduced this type of government. It was Christians, not Heathens, who paid divine honors to living kings. Heathens honored deceased kings. Equal rights of nature are mentioned in the paragraph, but not in relation to the Heathens.

14. B: Paine asserts that a monarchy is against the equal rights of nature and cites several parts of scripture that also denounce it. He doesn't say it is against the laws of nature. Because he uses scripture to further his argument, it is not despite scripture that he denounces the monarchy. Paine addresses the law by saying the courts also do not support a monarchical government.

15. A: To be *idolatrous* is to worship idols or heroes, in this case, kings. It is not defined as being deceitful. While idolatry is considered a sin, it is an example of a sin, not a synonym for it. Idolatry may have been considered illegal in some cultures, but it is not a definition for the term.

16. A: The essential meaning of the passage is that the Almighty, God, would disapprove of this type of government. While heaven is mentioned, it is done so to suggest that the monarchical government is irreverent, not that heaven isn't promised. God's disapproval is mentioned, not his punishment. The passage refers to the Jewish monarchy, which required both belief in God and kings.

17. A: The word *timorous* means being full of fear. The author concludes that extreme greed (avarice) makes people too afraid to be prosperous.

18. B: The author's attitude is closest to Choice *B, impassioned and critical*. Choice *A* is incorrect; the author is not *indifferent or fatigued*—on the contrary, there is a lot of energy and some underlying passion in the writing. Choice *C* is incorrect; the word *enchanted* means delighted, and the author is more critical and concerned of a monarchial government than enchanted with it. Choice *D* is not the best answer; although the author is passionate and critical of a monarchy, there is more logic than anger coming from the words.

19. B: To explain how monarchs came into existence and how Christians adapted to this way of government. Choice *A* is incorrect; the author does not agree that heathens were more advanced than Christians in this paragraph, it only explains the catalyst of the monarchial systems. Choice *C* is incorrect; the author would in fact disagree with the divination of the English monarchs. Choice *D* is incorrect; the paragraph *does* believe that monarchs cause damage, but the paragraph does not act as a counterargument to the one preceding it.

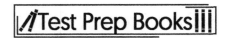
20. D: The author implies that the recorded history of creation is a collection, collage, or pattern taken from various accounts of cultures. Choice *A* is incorrect; there is no talk of cells or biology in the paragraph. Since "mosaic" modifies the word "account," we know that it is the account of creation that is mosaic, not creation itself, which makes Choice *B* incorrect. Choice *C* is also incorrect; the paragraph does not mention kings developing a system of recording accounts of creation.

Humanities

21. B: But in fact, there is not much substance to such speculation, and most anti-Stratfordian arguments can be refuted with a little background about Shakespeare's time and upbringing. The thesis is a statement that contains the author's topic and main idea. The main purpose of this article is to use historical evidence to provide counterarguments to anti-Stratfordians. Choice *A* is simply a definition; Choice *C* is a supporting detail, not a main idea; and Choice *D* represents an idea of anti-Stratfordians, not the author's opinion.

22. C: Rhetorical question. This requires readers to be familiar with different types of rhetorical devices. A rhetorical question is a question that is asked not to obtain an answer but to encourage readers to consider an issue more deeply.

23. B: By explaining grade school curriculum in Shakespeare's time. This question asks readers to refer to the organizational structure of the article and demonstrate understanding of how the author provides details to support the argument. This particular detail can be found in the second paragraph: "even though he did not attend university, grade school education in Shakespeare's time was actually quite rigorous."

24. A: Busy. This is a vocabulary question that can be answered using context clues. Other sentences in the paragraph describe London as "the most populous city in England" filled with "crowds of people," giving an image of a busy city full of people. Choice *B* is incorrect because London was in Shakespeare's home country, not a foreign one. Choice *C* is not mentioned in the passage. Choice *D* is not a good answer choice because the passage describes how London was a popular and important city, probably not an undeveloped one.

25. B: This sentence is an example of a metaphor. Metaphors make a comparison between two things, usually saying that one thing *is* another thing. Here, the author is saying that Shakespeare *is* "the great magician." Choice *A*, *personification*, is when an inanimate object is given human characteristics, so this is incorrect. Choice *C*, *simile*, is making a comparison between two things using *like* or *as*, so this is incorrect. Choice *D*, *allusion*, is an indirect reference to a place, person, or event that happened in the past, so this is also incorrect.

26. D: Remember from the first passage that anti-Stratfordians are those who believe that Shakespeare *did not* write the plays, so Stratfordians are people who believe that Shakespeare *did* write the plays. The author of Passage 2 is disbelieving and critical of the Stratfordian point of view. We see this especially in the second paragraph, where the author states it is a supposition "so wild that it can only be entertained by those who are prepared to accept it as a miracle." All of the other answer choices are incorrect.

27. D: Sentence 2 is a contrast to the idea in Sentence 1. In the first sentence, the author states that they, at one time, believed that Shakespeare was the author of his plays. The second sentence is a contrast to that statement by saying the author no longer believes that the author of the plays is Shakespeare. The other answer choices are incorrect.

28. B: This writing style is best described as persuasive. The author is trying to persuade the audience, with evidence, that Shakespeare actually wrote his own dramas. Choice *A*, expository writing, means to inform or explain. Expository writing usually does not set out to persuade the audience of something, only to inform them, so this choice is incorrect. Choice *C*, narrative writing, is used to tell a story, so this is incorrect. Choice *D*, descriptive writing, uses all five senses to paint a picture for the reader, so this choice is also incorrect.

29. C: Topography is the shape and features of the Earth. The author is implying here that whoever wrote Shakespeare's plays studied the physical features of foreign cities. Choices *A*, *B*, and *D* are incorrect. Choice *A* is simply known as climate. Choice *B* would just be considered the "agriculture of a particular area." Choice *D*, aspects of humans within society, would be known as *anthropology.*

30. A: The author of Passage 2 believes that Shakespeare the actor did not write the plays. We see this at the end of the first paragraph where the author contends that the "'Stratford rustic' is not the true Shakespeare." The author does believe that Shakespeare was an actor, as the author calls this Shakespeare a "Player" throughout the text, so Choices *B* and *D* are incorrect. Choice *C* is incorrect, as the author does not believe that Shakespeare wrote the plays.

Natural Science

31. C: *Extraneous* most nearly means *superfluous,* or *trivial.* Choice *A*, *indispensable,* is incorrect because it means the opposite of *extraneous.* Choice *B*, *bewildering,* means *confusing* and is not relevant to the context of the sentence. Finally, Choice *D* is incorrect because although the prefix of the word is the same, *ex-,* the word *exuberant* means *elated* or *enthusiastic,* and is irrelevant to the context of the sentence.

32. A: The author's purpose is to bring to light an alternative view on human perception by examining the role of technology in human understanding. This is a challenging question because the author's purpose is somewhat open-ended. The author concludes by stating that the questions regarding human perception and observation can be approached from many angles. Thus, the author does not seem to be attempting to prove one thing or another. Choice *B* is incorrect because we cannot know for certain whether the electron experiment is the latest discovery in astroparticle physics because no date is given. Choice *C* is a broad generalization that does not reflect accurately on the writer's views. While the author does appear to reflect on opposing views of human understanding (Choice *D*), the best answer is Choice *A.*

33. C: It presents a problem, explains the details of that problem, and then ends with more inquiry. The beginning of this paragraph literally "presents a conundrum," explains the problem of partial understanding, and then ends with more questions, or inquiry. There is no solution offered in this paragraph, making Choices *A* and *B* incorrect. Choice *D* is incorrect because the paragraph does not begin with a definition.

34. D: Looking back in the text, the author describes that classical philosophy holds that understanding can be reached by careful observation. This will not work if they are overly invested or biased in their pursuit. Choices *A*, *B*, and *C* are in no way related and are completely unnecessary. A specific theory is not necessary to understanding, according to classical philosophy mentioned by the author.

35. B: The electrons passed through both holes and then onto the plate. Choices *A* and *C* are incorrect because such movement is not mentioned at all in the text. In the passage, the author says that electrons that were physically observed appeared to pass through one hole or another. Remember, the electrons that were observed doing this were described as acting like particles. Therefore, Choice *D* is incorrect. Recall that the plate actually recorded electrons passing through both holes simultaneously and hitting the plate. This behavior, the electron activity that wasn't seen by humans, was characteristic of waves. Thus, Choice *B* is the correct answer.

36. C: The author mentions "gravity" to demonstrate an example of natural phenomena humans discovered and understood without the use of tools or machines. Choice *A* mirrors the language in the beginning of the paragraph but is incorrect in its intent. Choice *B* is incorrect; the paragraph mentions nothing of "not knowing the true nature of gravity." Choice *D* is incorrect as well. There is no mention of an "alternative solution" in this paragraph.

37. A: The important thing to keep in mind is that we must choose a scenario that best parallels, or is most similar to, the discovery of the experiment mentioned in the passage. The important aspects of the experiment can be summed up like so: humans directly observed one behavior of electrons and then through analyzing a tool (the plate that recorded electron hits), discovered that there was another electron behavior that could not be physically seen by human eyes. This summary best parallels the scenario in Choice *A*. Like Feynman, the colorblind person can observe one aspect of the world but through the special goggles (a tool), he is able to see a natural phenomenon that he could not physically see on his own. While Choice *D* is compelling, the x-ray helps humans see the broken bone, but it does not necessarily reveal that the bone is broken in the first place. The other choices do not parallel the scenario in question. Therefore, Choice *A* is the best choice.

38. B: The author would not agree that technology renders human observation irrelevant. Choice *A* is incorrect because much of the passage discusses how technology helps humans observe what cannot be seen with the naked eye; therefore, the author would agree with this statement. This line of reasoning is also why the author would agree with Choice *D*, making it incorrect as well. As indicated in the second paragraph, the author seems to think that humans create inventions and tools with the goal of studying phenomena more precisely. This indicates increased understanding as people recognize limitations and develop items to help bypass the limitations and learn. Therefore, Choice *C* is incorrect as well. Again, the author doesn't attempt to disprove or dismiss classical philosophy.

39. B: The word *conundrum* most nearly means *enigma*, which is a mystery or riddle. *Platitude* means a banal or overused saying. *Solution* is incorrect here; and *hypothesis* implies a theory or assumption, so this is also incorrect.

40. A: To prove to the audience the thesis of the passage by providing evidence suggesting that electrons behave differently when observed by the human eye. The thesis' best evidence is found in this paragraph because of the experiment depicting how electrons behave in a dual nature. Choice *B* is incorrect; the paragraph mentions nothing about the experiment being unethical. Choice *C* is incorrect; this is characteristic of the first paragraph. Choice *D* is incorrect; this is more characteristic of the second paragraph.

Science

1. C: There is a 99% probability of PCR testing identifying *Histoplasma*. GM assay was more specific for identifying *Aspergillus,* 95% to 85%. True positive is defined by sensitivity. The sensitivity of GM assay testing is less than 70%.

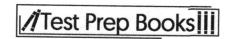

2. D: *Histoplasma* is detectable 90 days from exposure. PCR testing is able to detect *Histoplasma* 91 days from exposure—one day after sufficient organisms exist for detection. *Candida* is detectable 45 days from exposure. PCR testing is able to detect *Candida* 72 days from exposure—27 days after a sufficient number of organisms exist for detection. *Aspergillus* is detectable 118 days from exposure. ELISA testing is able to detect *Aspergillus* 134 days from exposure—16 days after a sufficient number of organisms exist for detection. *Candida* is detectable 45 days from exposure. GM assay testing is able to detect *Candida* 56 days from exposure—11 days after a sufficient number of organisms exist for detection.

3. B: The probability that the GM assay will identify *Candida* is 69%. Therefore, there's a 31% probability that it won't be identified. ELISA sensitivity and specificity for *Histoplasma* are both greater than 80%. False-negative probabilities are represented by the specificity of a given testing method. The sensitivity and specificity for GM assay testing for Aspergillus is 59% and 96% respectively. All testing methods had greater than 90% specificity for the organisms.

4. C: The sensitivity of PCR testing for *Histoplasma* is 99%, and the test can identify the organism one day after it reaches a detectable colony size. The sensitivity for GM assay testing for *Histoplasma* is 65%. If physicians rely on GM assay testing, they may determine that the individual doesn't have the *Histoplasma* infection. Treatment will depend on the presence or absence of the infection as indicated by testing. Waiting for PCR testing is based on the sensitivity of the test, not the individual's current symptoms. The subclinical phase of *Histoplasma* is 28 days.

5. A: ELISA testing detects *Candida* three days after the organism is present in sufficient numbers to be recognized. PCR detects the organism more than three weeks after it is first detectable. ELISA testing sensitivity for *Candida* is 87% and PCR testing is 92%. However, the ability to identify the presence of the organism earlier in the process of infection (allowing early intervention) outweighs the differences in the probability of identifying the presence of the organism. There's a 92% probability that PCR testing will identify the presence of *Candida*. PCR testing is more sensitive than ELISA: 92% versus 87%.

6. D: The main difference in the scientists' opinions is related to the cause of mad cow disease. The existence of species-specific proteins was used by Scientist 2 to support viral infection as the cause of the disease. Transmission rates of the disease and the conversion of normal proteins to prions were not debated in the passage.

7. A: Scientist 2 proposed that viruses were the cause of mad cow disease because chemicals inactivated the viruses. The remaining choices are correct.

8. B: According to Scientist 1, abnormal prions are capable of "refolding" normal proteins into harmful prions. Abnormal proteins accumulate to produce the damaging conglomerations. Scientist 2 didn't find species-specific DNA and used this fact to support viruses as the cause of mad cow disease. According to Scientist 1, prions are located in the central nervous system, not the peripheral nervous system.

9. D: Mad cow disease can be spread between animal species and from animals to humans through consumption of diseased animal products. The resulting damage to the central nervous system is irreversible and will eventually cause the death of the animal. Scientist 2 would not agree that the infecting agent contained amino acids, as they form proteins, and Scientist 2 believes that a virus causes the disease. Scientist 2 demonstrated that the infected tissue of animals that were infected by a different species didn't contain species-specific DNA, which would have been the expected outcome if the infecting agent were a protein.

10. B: The accumulated masses of abnormal prions eventually form sponge-like holes in the brain and spinal cord that result in death. The passage doesn't mention the effects of the synapses, nerves, or blood supply.

11. D: The absence of disease resulting from the inactivated viral particles best supports the views of Scientist 2. There were no species-specific DNA sequences found in the infected particles. Scientist 2 didn't support the existence of prions as the cause of mad cow disease.

12. C: The actual process of "refolding" the normal protein into the abnormal protein isn't clear from this passage. Scientist 1 claims that prions cause the disease. Prions are an abnormal protein, not a virus. Scientist 1 claims that mad cow disease is caused by abnormal proteins.

13. A: The main hypothesis for this study involved the influence of 2-AG levels combined with sleep deprivation on eating behaviors. The combination of the two conditions, not each one separately, constitutes the main hypothesis. The passage didn't discuss a placebo effect in the normal saline injection group.

14. D: The study results support the hypothesis because the participants who received 1-AG injections and were sleep deprived gained more weight than participants who received sterile normal saline injections. The remaining choices do not support the hypothesis.

15. A: Participant H gained more than 1 pound (450g) per day. There was little fluctuation in the day-to-day weight gain for each participant. Participant H in trial 2 gained more weight than participant D in trial 1.

16. B: Sleep deprivation increases the levels and duration of action of 2-AG, an endogenous cannabinoid, especially in the late afternoon. The stress effect increases with the degree of sleep deprivation. The passage doesn't discuss a relationship between sleep deprivation and the hunger response. Eating behaviors are increased in late afternoon as a result of the extended duration of 2-AG action.

17. A: Circadian fluctuations increase the levels of 2-AG during the afternoon and evening. This increase is believed to stimulate food intake beyond the point of satiety. Endogenous cannabinoids decrease gastric motility. 2-AG may have a calming effect on mood, but food intake is still increased in the presence of afternoon and evening levels of 2-AG. Endogenous cannabinoids work with the opioid system to mediate the pain response, not food-seeking behaviors.

18. A: In trial 3, the plants grown with the combined-wavelength LED's reached 150 millimeters by day 21. The plants grown with white light reached 160 millimeters by day 35. Therefore, the LED plants will be packaged 14 days sooner than the white light plants.

19. C: In trial 3, with LED lighting that included green and yellow wavelengths, plant growth was greater than trial 1 or trial 2 with either blue or red wavelengths. However, from the available information, it can only be said that green and yellow wavelengths *contributed to* plant growth in trial 3, but not that green and yellow wavelengths *alone* were responsible for plant growth in trial 3. There was plant growth in all lighting conditions.

20. B: In trial 1, from day 28 to day 35, white light growth increased by 71 millimeters, and red light increased by 78 millimeters. In trial 2, from day 28 to day 35, white light growth increased by 71 millimeters, and blue light increased by 78 millimeters.

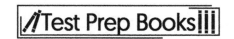

21. C: The average daily growth with LED lighting was not twice the white light average daily growth. LED systems did result in better growth rates and they do require less water and electricity. However, the question is based on recorded average daily growth, and that rate was not double the white light rate.

22. D: The passage says that green and yellow wavelengths are reflected by the plant. Therefore, it's expected that those wavelengths would result in slower growth than the blue or red wavelengths, which are absorbed.

23. B: *Anacardiaceae Mangifera* is the family and genus name for the mango. The *Colletotrichum* fungus causes the spoilage. The remaining choices are correct.

24. C: According to Table 1, at 20 days, the fungal level at 7.5 °C was the same as the fungal level at 5 °C. Because the 7.5 °C temperature is less expensive than the 5 °C temperature, the 7.5 °C model is best. The 10 °C model is less expensive than the 7.5 °C, but fungal levels are greater. Only the 7.5 °C and 5 °C models had acceptable fungal levels at 20 days.

25. B: The hot water wash pre-treatment fungal level increased by 15 millimeters from day 26 to day 28 at 10 °C. It was the single largest one-day increase across the trials. Air cooling at 10 °C increased by 5 millimeters from day 14 to day 16. Air cooling at 7.5 °C increased by 7 millimeters from day 26 to day 28. Hot water wash at 7.5 °C increased 11 millimeters from day 26 to day 28.

26. C: The hot water wash fungal level at 10 °C reached 35 millimeters on day 20. The maximum fungal level for air cooling at 5 °C was 19 millimeters, and at 10 °C, 35 millimeters on day 26. The maximum fungal level for hot water at 7.5 °C was 39 millimeters on day 28.

27. C: The fungal levels were acceptable with air cooling at 7.5 °C, and the 7.5 °C temperature is less expensive to maintain. Air cooling and hot water wash pre-treatment at 5 °C resulted in acceptable fungal levels, but the 5 °C temperature is costlier. Fungal levels were not acceptable at 28 days at 10 °C.

28. B: Shipping mangoes at 5 °C is costly, but for the 28-day trip, the fungal levels were only acceptable in the 5 °C model. Air cooling fungal rates at 5 °C were lower than the hot water wash rates, but each was acceptable. Fungal rates at 7.5 °C and 10 °C were unacceptable.

29. A: The correlation was positive, because when one variable increased, the other increased, and when one variable decreased, the other decreased. There are two forms for a curvilinear relationship. In one curvilinear relationship, when variable 1 increases, a second variable increases as well, but only to a certain point, and then variable 2 decreases as variable 1 continues to increase. In the other form, variable 1 increases while variable 2 decreases to a certain point, after which both variables increase. In a negative relationship, high values for one variable are associated with low values for the second variable.

30. D: Mitochondrial activity is suppressed by elevated blood glucose levels. Mitochondria use sugar to produce cellular energy, and the presence of large numbers of mitochondria in BAT gives BAT a brownish color. Mitochondria contain DNA and can reproduce additional mitochondria when additional energy is required. In the body, mature red blood cells are the only cells that don't contain mitochondria.

31. D: Blood sugars for all groups identified in Figure 5 were highest at 2 p.m.

32. B: Circadian rhythms control sleep by stimulating the release of melatonin from the pineal gland. Ganglia are nerve cells, not hormones, that affect sleep. BAT doesn't release sugar; it utilizes sugar for heat production. Shivering on cold mornings is a desirable form of thermogenesis but isn't associated with sleep.

33. D: The average blood sugar for participant 6 was 115. Participant 1 was 105, participant 3 was 112, and participant 4 was 82.

34. B: The daytime blood glucose levels in Figure 1 decreased as the day progressed. The blood glucose levels in Figure 5 peaked for the day at 2 p.m. Night blood glucose levels didn't reach 100. Group I's levels are irrelevant to the question.

35. A: Based only on Figure 2, the researchers' hypothesis wasn't confirmed. Subsequent trials confirmed the hypothesis.

36. A: Increased algae levels can block sunlight, limiting growth of species inhabiting lower zones. The passage doesn't identify the effects of rainfall or cold temperatures on phytoplankton growth, so Choices *B* and *D* are incorrect. The passage identifies the effect of phosphorus and nitrogen residue on algae growth, but not as a food source for cyanobacteria.

37. D: The passage identifies freshwater contamination by phosphorus and nitrogen as the most common cause of algae overgrowth. Population density would be more common in Florida than Kansas.

38. B: As algae levels increase above normal, organisms in the aphotic level plants don't receive adequate light for normal growth and oxygen levels are decreased, resulting in the death of oxygen-dependent species.

39. C: Algae block the sunlight, which limits growth.

40. A: Algae growth was greater in July, which limited the amount of light reaching the lower zones of the lake, decreasing the levels of cyanobacteria. Cyanobacteria existed in less-than-normal concentrations at 20 meters, but there were measurable levels of the organisms. Algae growth at 3 meters wasn't measured. The passage states that cyanobacteria growth is associated with algae growth, not independent of algae growth.

Dear ACT Test Taker,

We would like to start by thanking you for purchasing this practice test book for your ACT exam. We hope that we exceeded your expectations.

We strive to make our practice questions as similar as possible to what you will encounter on test day. With that being said, if you found something that you feel was not up to your standards, please send us an email and let us know.

We would also like to let you know about other books in our catalog that may interest you.

SAT

This can be found on Amazon: amazon.com/dp/1628458984

ACT

amazon.com/dp/1628458844

ACCUPLACER

amazon.com/dp/162845945X

TSI

amazon.com/dp/162845721X

SAT Math 1

amazon.com/dp/1628454717

AP Biology

amazon.com/dp/1628456221

We have study guides in a wide variety of fields. If the one you are looking for isn't listed above, then try searching for it on Amazon or send us an email.

Thanks Again and Happy Testing!
Product Development Team
info@studyguideteam.com

FREE Test Taking Tips DVD Offer

To help us better serve you, we have developed a Test Taking Tips DVD that we would like to give you for FREE. **This DVD covers world-class test taking tips that you can use to be even more successful when you are taking your test.**

All that we ask is that you email us your feedback about your study guide. Please let us know what you thought about it – whether that is good, bad or indifferent.

To get your **FREE Test Taking Tips DVD**, email freedvd@studyguideteam.com with "FREE DVD" in the subject line and the following information in the body of the email:

     a. The title of your study guide.

     b. Your product rating on a scale of 1-5, with 5 being the highest rating.

     c. Your feedback about the study guide. What did you think of it?

     d. Your full name and shipping address to send your free DVD.

If you have any questions or concerns, please don't hesitate to contact us at freedvd@studyguideteam.com.

Thanks again!

Made in the USA
Columbia, SC
10 August 2021